In Defence of Science
Science, Technology, and Politics in Modern Society

Science holds a central role in the modern world, yet its complex interrelationships with nature, technology, and politics are often misunderstood or seen from a false perspective. In a series of essays that make extensive use of original work by sociologists, historians, and philosophers of science, J.W. Grove explores the roles and relationships of science in modern technological society.

Modern science can be viewed from four related perspectives. It is an expression of human curiosity – a passion to understand the natural world: what it is made of, how it is put together, and how it works. It is a body of practice – a set of ways of finding out that distinguish it from other realms of inquiry. It is a profession – a body of men and women owing allegiance to the pursuit of knowledge – and for those people, a career. And it is a prescriptive enterprise in that the increase of scientific understanding makes it possible to put nature to use in new kinds of technology.

Each of these aspects of science is today the focus of critical scrutiny and, often, outright hostility. With many examples, Grove exposes the threats to science today: its identification with technology, its subordination to the state, the false claims made in its name, and the popular intellectual forces that seek to denigrate it as a source of human understanding and progress.

J.W. GROVE is Professor of Political Studies (Emeritus), Queen's University. Among his other books are *Government and Industry in Britain* and *Organised Medicine in Ontario*.

J.W. GROVE

In Defence of Science:
Science, technology, and politics in modern society

UNIVERSITY OF TORONTO PRESS
Toronto Buffalo London

Toronto Buffalo London
Printed in Canada

ISBN 0-8020-2634-6 (cloth)
ISBN 0-8020-6720-4 (paper)

Printed on acid-free paper

Canadian Cataloguing in Publication Data

Grove, J.W. (Jack William), 1920-
In defence of science

Includes bibliographical references and index.
ISBN 0-8020-2634-6 (bound) ISBN 0-8020-6720-4 (pbk.)

1. Science – Social aspects. 2. Science and state. I. Title.

Q175.5.G76 1989 306′45 C88-094888-4

This book has been published with the assistance of the Canada Council and the Ontario Arts Council under their block grant programs.

To John Meisel

Wise, fair-spoken, and persuading

Contents

Preface

Each chapter in this book is, in essence, a free-standing essay. Each treats of an aspect of the complex mutual relationships that exist between science and nature, science and technology, science and politics, and politics and nature. Scientists study nature; but nature places constraints on what they can discover about it. Scientific knowledge is often useful and thus feeds technology; but technology, in turn, affects the practice of science, for example by making possible new techniques and instruments. Science impinges on politics when advances in knowledge pose questions for public policy; and politics impinges on science because governments today seek to sponsor and promote scientific work 'in the national interest' and control its direction. Ideas about the social and political order can affect the manner in which scientists interpret the natural world. Ideas about nature – which are not always scientific ideas and are sometimes opposed to them – may raise political controversies. For example, fundamentalist religious ideas about the natural world run directly counter to scientific ideas, and fundamentalists seek explicitly political means to attack and discredit them.

This book makes extensive use of original work by sociologists, historians, and philosophers of science, professional students of politics, and scientists writing for a lay public. It is often sharply critical of much that has been written and said of late about science and its relation to the wider society in which it is practised.

Acknowledgments

Parts of chapter 2 first appeared in *Minerva: A Review of Science, Learning and Policy* 18, no. 2 (summer 1980), 293–312. Chapter 5 is a revised and expanded version of an essay published in the same journal, vol. 23, no. 2 (summer 1985), 216–40. I wish to thank the editor of *Minerva*, Professor Edward Shils, and the publishers, the International Council on the Future of the University, for permission to use this material.

I am indebted to Bernice Gallagher (occasionally, ably assisted by Leisa Macdonald and Frances Shepherd) for typing early drafts of the book, and to Shirley Fraser for preparing the final manuscript for publication.

IN DEFENCE OF SCIENCE

1

Science as Understanding

In Nature's infinite book of secrecy
A little I can read ...
Anthony and Cleopatra, I, ii, 11–12

The natural sciences constitute a mode of thought and practice through which the world is experienced and made comprehensible. The primary *motive* for doing science is curiosity and a passion to know. The fundamental *aim* of science is to increase our understanding of the world: what it is made of, how it is put together, and how it works. In this sense, all science is cosmology and all the separate sciences knit together in a seamless web. Writing half a century ago, J.W.N. Sullivan said: 'Knowledge for the sake of knowledge, as the history of science proves, is an aim with an irresistible fascination ... which needs no defence. The mere fact that science does, to a great extent, gratify our intellectual curiosity is a sufficient reason for its existence ... For science to have inspired such ardour and devotion ... it is obvious that it must meet one of the deepest needs in human nature' (1933, 1938: 214–15). The response of modern cynics is, of course, to exclaim 'But what an arrogantly elitist view! You mean science has an irresistible fascination for some; that it meets the deepest needs in *their* nature.' This argument is not compelling. All 'high cultural' activities are 'elitist' in precisely the same sense. If this were not so it would have been unnecessary for the mid-twentieth century to have invented the notion of 'pop culture.'

The greatest of the nineteenth-century expositors of science for a wider interested public, the biologist Thomas Henry Huxley (1826–95), once said that 'in the ultimate analysis everything is incomprehensible, and the whole object of science is simply to reduce the fundamental incomprehensibilities to the smallest number.' Science has been remarkably successful in pursuing this aim. Only the most perverse could possibly deny that our understanding of the world and of our own place in it is incomparably clearer, deeper, and more detailed than it was five hundred, one hundred, or even as little as fifty years ago. Science makes progress, it 'adds to our knowledge'; but it does so not so much by a linear process of accumulating 'facts' about the world as by a continuous reinterpretation and development of earlier ideas. Science is a kind of story-telling, like the making of myths, but it differs from myth-making in important respects, notably in that its stories are constantly being rewritten in order to conform to the way the world is discovered to be. Free speculation is checked by close and critical examination of the results of observation and experiment.

Some of the most famous myths known to us from ancient times are extremely complex and ingenious intellectual structures, as indeed are many of those found in the contemporary or near contemporary tribal societies that have been investigated by social anthropologists. In one of their aspects these myths are primitive cosmologies. Like science they give expression to the universal human need to find order and regularity in the world and to resolve intellectual problems arising out of Man's encounter with Nature. Unlike science, however, they tend to be tenaciously guarded against criticism and change. It is essentially the spirit of unfettered and, above all, critical inquiry that has marked science off from myth. Scientific inquiry begins, as Sir Peter Medawar said, 'as a story about a possible world – a story which we invent and criticise and modify as we go along'; but it must end by being 'as nearly as we can make it, a story about real life' (1969: 59). François Jacob, another biologist, expressed the same thought: 'For science, there are many possible worlds; but the interesting one is the world that exists and has already shown itself to be at work for a long time. Science attempts to confront the possible with the actual' (1982: 12).

SCIENCE AND REALITY

The belief in an external world independent of the perceiving subject is the basis of all natural science.

Albert Einstein

With some important exceptions that will be discussed later, scientists are realists; that is to say, they adhere, whether explicitly or not, to the philosophical doctrine known as Realism. Realism asserts, first, that there is an external world that exists independently of us, a world that existed before there were any 'knowing subjects,' and that is likely to exist long after we have vanished from the scene; secondly, that this world is knowable (albeit imperfectly) by us, as it actually is and not merely as appearance; and lastly, that it is unaffected by our ideas about it[1] or, more specifically, that we cannot change it merely by thinking and speaking about it. Even philosophical idealists, such as Bishop Berkeley (1685–1753), who hold that nothing exists except minds and mental states, must believe in a world external to themselves to the extent that they assume the existence of minds other than their own.[2] Only solipsists believe that nothing is real except themselves and *their* mental states.[3] The issue, then, that divides Realism from its contradictory doctrine,[4] Idealism, is not so much the existence of an external world as such, but rather, the question of how we can know it. Idealists ask of the realist how it can be known (by which is meant, known for certain) that a world independent of minds exists and, even if it does, how the realist can claim to know it as it actually is. That is to say, how can these realist claims be *justified*? The answer is that they cannot be justified in the sense of being demonstrated to be conclusively true; but neither can the counter-claims of the idealists. Both Realism and Idealism are metaphysical doctrines. They cannot be proved, but neither can they be refuted. But both can be discussed and criticized.

Few modern scientists would want to say that the world is knowable definitively. The experience of science is that ideas (that is, theories) about the way the world is constituted are constantly subject to change in the light of new discoveries, observations, and experimental results. The entire history of the scientific enterprise attests to the fact that very little can be taken as established or warranted truth – that is, known for certain. This is not to say that what is conjectured to be true may not in fact *be true*; it is to say that we cannot know for certain whether it is or not. This may appear to some as a very unsatisfactory state of affairs, which is understandable given what seems to be a human longing for the stable and the certain, but it in no way supports Idealism. For how could it be possible for our ideas about the world to be mistaken if the world did not exist independently of our minds or if there were no way of gaining privileged access to it? We need to distinguish between what we know, or think we know, about the world and the concept of reality itself. The fact that our knowledge of the world is tentative and imperfect is no reason for supposing that reality exists only in our minds. Though

Realism is a metaphysical doctrine, there are plausible empirical and logical arguments that support it.

Immanuel Kant (1724–1804) argued that it is we who impose order on the world ('Our intellect does not draw its laws *from* nature, but imposes it laws *upon* nature' – *Prolegomena*, sec. 37). He was right in supposing that we are not passive receptacles for knowledge that flows into our minds *from* reality,[5] but that in an important sense we make reality; but it would be wrong to assume from this that reality is *merely* socially constructed as many contemporary philosophers and sociologists of science contend. For there are at least intimations that there is something 'out there' that leads us to impose order on nature. As Sir Peter Medawar argued, 'There is *something* in the fact that the atomic weights of the elements are very nearly whole numbers' (emphasis his). Scientists propose theories about the nature of the world that frequently predict the existence of things that have not yet been detected. One of the most dramatic examples of this is Mendeleev's periodic table of the elements. What Mendeleev showed in 1869 was that if the elements are arranged according to their atomic weights they exhibit a periodicity of properties. For example, fluorine, bromine, chlorine, and iodine all have some properties in common that differ from those of other groups of elements such as sodium, potassium, and calcium. He was able, from gaps in his table, to predict the existence of other, unknown elements and these were all subsequently discovered. Further, the fact that it was later discovered that the atomic weights of the elements are not quite whole numbers led to the discovery of isotopes and, indeed, contributed to the development of modern ideas about atomic structure. There are many examples of entities predicted by theory the existence of which was later confirmed. Examples are the blood capillaries, the gene, viruses, and even atoms, which have now been photographed.[6]

Theories precede discoveries. Many of the physicists' 'fundamental' particles such as the positron (a positively charged electron) are introduced as theoretical entities because they are entailed by a specific theory, and are later detected experimentally. Such particles are not, of course, directly observable (they are only seen as traces in bubble-chamber photographs), but it would be odd not to accept them, and in practice most physicists do accept them, as real. In fact, the distinction between what is said to be directly observable and what is indirectly observable is spurious and belongs to an outmoded epistemology. All observations have the character of hypotheses, and all observational terms – for example, 'the sun' – embody some theory about what is currently observed.

Nothing is directly observable (see below).[7] It is true that the reality that is discovered by the physicists (but physics is not the whole of science) is sometimes very strange. For example, high-energy physicists convert energy into matter by accelerating particles (such as protons) in giant accelerators up to velocities very close to the velocity of light (more than 99 per cent in some cases) and then bringing them to an abrupt halt. The increased mass that the particles have thus acquired (in accordance with – that is to say, corroborating – Einstein's famous prediction that $e = mc^2$) is 'stripped off' in a shower of 'new' particles of matter and anti-matter. These must be real in some significant sense even though most of them are not present in ordinary matter (which consists essentially of only three kinds of entities, – electrons, protons, and neutrons), and they exist in the apparatus for no more than fractions of a millionth of a second. But, again, the fact that reality is strange is no reason for thinking that it is an illusion.

When Dr Johnson was asked by Boswell how he would refute Idealism, he replied: 'Sir, I refute it thus' – and kicked a large stone 'with mighty force ... 'til he rebounded from it.' Similarly, the reality of an electrostatic charge is painfully obvious to every homeowner who possesses central heating, metal door handles, wall-to-wall carpeting, and an inefficient humidifier. Indeed, Karl Popper (who is probably the first realist to have used Dr Johnson as an exemplar) has roughly defined as 'real' anything that we can affect and that, in turn, can have an effect on us. This is a fruitful definition, for it enables us to say that intangible things can also be real, for example ideas and theories. These are real in the important sense that, once created, they exist independently of us: they can affect us and we can affect them (for example, by criticizing, changing, and perhaps rejecting, them).[8] Reality need not be confined to material objects. The definition, incidentally, also clears up part of the confusion surrounding the 'directly observable' and the 'indirectly observable.' Many nineteenth-century positivists, like the physicist-philosopher Ernst Mach (1838–1916) for example, rejected forces such as gravitational force as 'occult' (not real) on the grounds that they are not directly observable. But since gravitational force acts on us, and we can act on it (for instance, by generating sufficient energy to counteract it locally, as in the space shuttle), there is no reason to exclude it as real.

The empirical confirming of entities that are predicted by scientific conjectures about the nature of the world is of some comfort to realists, but much more important is the fact that science, like life in general, is full of surprises. We learn most when our predictions turn out *not* to be correct, that is to say, when we make mistakes. Scientists 'get in touch with reality' by becoming aware of the existence of things and states of affairs that they did not previously suppose were there. Kicking stones is all very well, but we learn more when nature kicks back. A nasty pain in the foot is a confirmation of the reality of stones, but we would learn something much more interesting about the nature of the world and ourselves if we kicked a stone and felt nothing.

The scientist's brand of realism asserts that we try to approach the identification and description of reality through a continuous process of trial and error. It is not, therefore, an a priori ontological realism of the sort espoused by certain realist philosophers who would assert, for example, that we must first know that objects exist before we can say that 'this table' exists. On the contrary, it is a realism that asserts that we only know (or more accurately, conjecture) that objects exist because we know (conjecture) that 'this table' exists – and have good reasons for doing so. Scientific realism is, in a phrase used by Roger Trigg, 'a neutral doctrine about the *character* of what is real' (1980: xix; emphasis mine). Martin Gardner once characterized science as 'a complicated game in which the universe seems to possess an uncanny kind of order, an order that it is possible for humans to discover in part, but not easily.' Scientists put questions to nature and they get comprehensible replies back; but the process is rather like a conversation in a dark room; one has to infer what the room contains and what its inhabitants are doing from partial answers to questions that are largely guesses about what may be the case.

SCIENCE AND CERTAINTY

To know and to be certain are the same thing ... what comes short of certainty, I think, cannot be called knowledge.

John Locke, Letter to Stillingfleet, 1697[9]

At one time, chiefly under the influence of the immense success of classical, Newtonian mechanics, but partly also under the influence of defective notions of scientific method deriving from Francis Bacon (1561–1626), science was commonly conceived as a steady accumulation of certain knowledge. Everyone of any consequence eventually believed that Newton's laws were immutable, even the sceptic David Hume, who predicted that they would 'probably go down triumphant to the latest posterity.' There developed an enormous faith in the infallibility of science; in the notion that 'genuine' scientific statements must be unchallengeably true, that through science we could arrive at certain knowledge of the world. Science was thought to be cumulative, patiently uncovering truth and correcting earlier mistakes, always moving towards some finally completed edifice, until eventually – though no doubt far into the future – we would know everything. This view was laid to rest (though it has not, even now, penetrated totally into the popular mind) with the revolution wrought in the early part of this century by Einstein, and the dethroning of classical mechanics by the new quantum physics. Its abandonment led many philosophers – of whom Edmund Husserl (1859–1938) was a leading representative[10] – to despair of science as the intellectual basis of modern civilization, to attack it, and to call for a return to the ancient search for certain knowledge through pure reason and intuition. That knowledge acquired by science is not, in any final sense, certain should not trouble realists, for if we hold that the world exists independently of us, we necessarily accept that it is not our creation, and if we accept that we must also accept that our ideas about it may be mistaken.

The uncertainty of science is in part accounted for by the fact that it is not based on 'hard facts'; that is to say, ultimately on sense experiences, as the old empiricist-phenomenalist philosophy of science assumed. Science starts from problems. It is, in Judson's phrase (1980), a *search for solutions*, solutions to non-trivial problems; or, as Medawar said (1967), it is the art of the soluble and important. Good scientists study the most important problems they think they can solve.[11] No scientist, said Medawar, is admired for failing in the attempt to solve problems that lie beyond his competence. Scientific research consists in tackling problems, whether experimental or theoretical, and proposing conjectural solutions to them. It is impossible for anyone (let alone a scientist) to start from observations because it is impossible to observe anything in the absence of some idea of what we are looking for; we need some criteria of relevance. 'Why can't you draw what you see?' Medawar wrote: '[This] is the immemorial cry of the teacher to the student looking down the microscope for the first time at some quite unfamiliar preparation he is called upon to draw. The teacher has forgotten, and the student himself will soon forget, that what he sees conveys no information until he knows beforehand the kind of thing he is expected to see' (1967: 133).

This fact was long ago recognized by Charles Darwin. Writing to Henry Fawcett[12] (18 September 1861) Darwin said: 'How odd it is that anyone should not see that all observation must be for or against some view if it is to be of any service'; and to Sir Charles Lyell[13] (1 June 1860), referring to a critic of Darwin's *Origin of Species*, 'On his standard of proof, natural science would never progress, for without the making of theories I am convinced there would be no observation.' The necessity for theory to precede observation was recognized even earlier by Auguste Comte (1798–1857), who wrote in his *Positive Philosophy* (1829): 'Our mind needs a theory in order to make observations. If in contemplating the phenomena we do not link them immediately with some principles, it would not only be impossible to ... draw any useful conclusions, we should be unable to remember them, and, for the most part, the facts would not be noticed by our eyes.' This observation seemed to Comte to pose the question: Where, then, did the first scientific theories come from? His answer was that they were preceded by 'theological conceptions.' This idea has something in common with Karl Popper's contention that scientific theorizing originated in the making of rational myths, that is, in myths that were criticizable and open to rejection after rational discussion.

The research process originates in what the American philosopher John Dewey called 'the occurrence of a felt difficulty' – that is, in a problem-situation. A theory (or hypothesis)[14] is then proposed to resolve the problem. Thinking is not merely a matter of stimulus and response. We react to situations, it is true, and in that sense 'start from facts,' but we also anticipate situations, that is, theorize about their possibility. At the same time, all theorizing aimed at solving a problem is theorizing in the light of background knowledge. This background knowledge leads us to possess expectations. A sudden noise in the night generates a hypothesis, 'there's a burglar downstairs,' but only because we have certain background knowledge (or prior expectations), for example, that it is normally quiet in the house at night. Once proposed as a solution to the problem, the theory must be tested; in this simple example, go and look (it may be the cat). But if science (or indeed any serious intellectual inquiry) is to make progress, a theory must be independently testable (or at least criticizable); it must not merely 'explain the facts before it'; if it does only that it is simply ad hoc, confirmed by the facts it explains. Ad hoc theories may be adequate, for the most part, in everyday life; but, in science, theories must also be able to account for new observations whose possibility is deducible from those theories. The aim of science, as Popper has consistently argued, is not to accumulate observations but to replace less good theories with better ones, such as those with greater explanatory power. As Einstein said in his autobiography: 'A theory is the more impressive ... the more various kinds of things it relates, and the more far-reaching is its area of applicability.'

Every solution to a problem in science generates new problems (sometimes a host of problems). One of this century's great life scientists, Salvador Luria, has said: 'Everyone knows that in [scientific] research there are no final answers, only insights that enable us to formulate new questions' (1984: 113). Another distinguished life scientist, Lewis Thomas, has written: 'The only solid piece of scientific truth about which I feel totally confident is that we are profoundly ignorant about nature ... It is this sudden confrontation with the depth and scope of [our] ignorance that represents the most significant contribution of 20th century science to the human intellect ... Now that we have begun exploring in earnest [in biology], doing serious science, we are getting glimpses of how huge the questions are' (1979: 73). It might be argued that this picture of science as a quest for an understanding of the world that is at once progressive and uncompletable – moving closer to a truth that it can never reach – is an illusion that stems from the relative immaturity of sciences like biology (which is arguably in its infancy compared to physics). Certainly, in those areas of modern physics that are devoted to uncovering the deepest mysteries of the universe – theoretical cosmology and experimental particle physics – which seek to explain the very large in terms of the very small, the claim is now being made that the end of the road may indeed be in sight. Specifically, Einstein's dream of a grand unifying theory that would account for all the fundamental forces that govern the entire universe, and thus explain why the universe must be constituted as it is, is said to be attainable in the course of the next few decades.[15]

It would be well, however, to be cautious of all such claims. As we have seen, 'the end of physics' has been forecast before, and one recent writer (a physicist) has cautioned that 'We may look back on our current attempts to understand the origins of the universe as hopelessly inadequate, like the attempts of medieval philosophers trying to understand the solar system before the revelations of Kepler, Galileo and Newton ... But it is also possible that we are near the end of our search. No one knows' (Pagels 1985). That is the ultimate uncertainty in science: it is impossible to predict what will be discovered next. But even if that part of physics did reach the end it would not be the end of science (or of physics). There would be left, as Martin Gardner has ironically observed, 'such trifles as explaining how the basic forces and particles manage to get together and write books about themselves' (1985: 31). The conviction that science is uncertain, perhaps better expressed as the principle that scientists should avoid dogmatism, came about largely as a consequence of the overthrow of classical physics. This event prompted Sir Arthur Eddington to write, in *The Nature of the Physical World* (1929): 'If the scheme of philosophy which we now rear on the scientific advances of Einstein, Bohr, Rutherford and others is doomed to fall ... it is not to be laid to their charge that we have gone astray. Like the systems of Euclid, of Ptolemy, of Newton, which have served their turn, so the systems of Heisenberg and Einstein may give way to some fuller realisation of the world. But in each revolution of scientific thought new words are set to the old music ... Amid all our faulty attempts at expression, the kernel of scientific truth steadily grows.' Einstein himself wrote (in a letter to a friend in 1917): 'No matter how we may single out a complex from nature ... its theoretical treatment will never prove to be ultimately conclusive ... I believe that this process of deepening of theory has no limits.'

It can be shown, as a matter of simple logic, that the advancement of knowledge is incompatible with certainty. If the growth of knowledge means that scientists operate with theories of increasing content, then it must be that they operate with theories of *decreasing probability*. Any statement that says more (has greater empirical content) than another statement will be *less* probable than the second statement because it is more likely to be false. 'It will rain on Friday and be fine on Saturday' is less probable than 'It will rain on Friday.'

Herbert Simon, in his Karl Compton Lectures (1968: 23–4), has a beautiful story that can be applied to the progress of science and its relationship to the world.

> We watch an ant making his laborious way across a wind-and-wave-moulded beach. He moves ahead, angling to the right to ease his climb up a steep dunelet, detours round a pebble ... Thus he makes his wearing, halting way ... I sketch the path on a piece of paper. It is a sequence of irregular, angular segments, not quite a random walk, for it has a sense of direction, of aiming toward a goal. I show the sketch to a friend. Whose path is it? An expert skier, perhaps, slaloming down a steep and somewhat rocky slope. Or a sloop beating upwind in a channel dotted with islands and shoals. Perhaps it is a path in a more abstract space: the course of search of a student seeking the proof of a theorem in geometry. Whoever made the path, and in whatever space, why is it not straight? Why does it not aim directly from its starting point to its goal?

The answer is, of course, that whoever made the path had a general sense of where the goal lay, but could not foresee all the obstacles (problems) that lay between. The scientist's path, like that of the ant's is irregular, complex, and hard to describe. But the complexity of the two paths reflects a complexity in nature rather than complexity in the scientist or in the ant. The theory of knowledge that is the accompaniment of scientific realism is an *active* theory. As against the *passive* theory of knowledge, which has knowledge 'coming to us' or 'given in experience,' the active theory looks upon the pursuit of knowledge as a process of discovery, a walk over a difficult terrain bringing the ever-increasing understanding that comes about by going out and looking, by seeking for the truth.

SCIENCE AND COMMON SENSE

> Science is, I believe, nothing but trained and organised common sense ... The man of science, in fact, simply uses with scrupulous exactness the methods which we all, habitually and at every moment, use carelessly.
>
> Thomas Henry Huxley

Huxley was concerned to defend science in what was for much of his lifetime (1825–95) a hostile social environment, and he chose to do so by pointing to its essential ordinariness. Science was simply the rational method writ large; its conclusions were those to which all honest and inquiring minds must ultimately come to assent. His defence of science against its enemies was to argue that it was different from common sense only with respect to its rigour and its precision. To a degree this is true: to the extent to which in science, as in ordinary life, we learn by experience, by trial, and (especially) error. (As Oscar Wilde said, experience is the name everyone gives to his mistakes.)[16] Nevertheless, science can never be satisfied with common sense; indeed, it is through long trial and error that scientists have learned to be sceptical of common sense. It is true that there is a link between science and common sense in the historical relationship that has existed in the accumulation, over several millennia, of practical lore such as woodcraft, the knowledge of the naturalist, and the making of star maps. There are links also in the fact that scientific explanations are often directed to the correction and improvement of common-sense explanations; and in the fact that much of today's 'common sense' is yesterday's science (few people today believe that the earth is flat or that malaria is due to breathing foul air). There are, nevertheless, equally obvious and significant differences.

Common sense may accept things as self-evident for no other reasons than authority or habit. It is not as rational as Huxley would have had us believe. This fact is not surprising, for as Alfred North Whitehead once said, it requires an unusual mind to undertake an analysis of the obvious. Many of the most imaginative flights of scientific creativity fly in the face of the 'evidence' – that is, the evidence of the senses or ordinary reasoning. It is surely common sense on both these grounds to suppose that the length of a body and time as measured by a clock are invariant with respect to the state of motion of the observer. Yet Einstein showed why this must be false; and observations with extremely delicate instrumentation later showed him to be right. It is self-evident to most people who have never heard of a Möbius strip (a figure that can easily be produced by twisting a strip of paper once and glueing the ends together) that every surface must have two sides. But this too is false. Again, most people believe that a stone held shoulder high by a person walking briskly forward and then dropped will fall straight down. In fact it will move forward as it falls, in accordance with Newtonian mechanics. This kind of 'intuitive physics,' as McCloskey has called it (1983), is quite common. It is common sense to believe that a moving body with a smooth surface will set up less wind resistance than one with a rough surface. But this apparently does not apply to golf balls, which is why they are nowadays manufactured with dimples. The reason was given by R.V. Jones in the 6th Lindemann-Simon Lecture at Oxford University in 1965.[17]

All of this, however, is more or less trivial. What is important is that a theory of knowledge that seeks to ground knowing in simple intuition and the evidence of the senses is bound to fail. It has to be accepted that there is nothing immediate about 'experience' and therefore nothing unchallengeable about it either. The psychological feeling of certainty that we possess about common-sense evidence, as the philosopher Morris Cohen remarked, is often nothing more than an inability to conceive the opposite of what we happen to believe. Common sense constantly tends to 'elevate the plausible into the axiomatic' (Cohen 1964: 48 and 83). It can be said, indeed, that the progress of scientific thought largely depends upon the capacity of the mind to function with concepts that are far removed from everyday common-sense experience. As Galileo wrote of Copernicus: 'I cannot express strongly enough my unbounded admiration for the greatness of mind of [this man] who conceived the heliocentric system and held it to be true ... in violent opposition to the evidence of his senses.' Such a transcending of what seems plausible at the moment is not easy to achieve, even in the practice of science.[18]

A good example can be drawn from Eddington's investigations, in the 1920s, of the companion star to Sirius, the brightest star in the Southern Hemisphere. This companion star, Sirius B, was calculated to have about 95 per cent of the mass of the sun, yet to be only slightly larger than the earth, and to have a much hotter surface than that of the sun which should make it extremely bright rather than, as it is, very faint. It is, in fact, a collapsed star of a type now known as a white dwarf. Eddington supposed that a collapsing star might go on collapsing until its mass would be compressed into a diameter of a few kilometres at most (now known as a neutron star). Eventually, its gravitational pull would become so strong that no light or other radiation could escape from it (it would then be a black hole). At that point, as he put it, 'the star can at last find peace.' But he refused to accept this result. Various accidents, he said, might intervene to save the star, but he wanted more protection than that. He thought that there ought to be a law of nature to prevent so absurd an outcome. (In fact there is such a law – or at least a physical constant, now known as Chandrasekhar's Limit – that does prevent some stars from so behaving.)

REALISM, QUANTUM PHYSICS, AND THE ANTHROPIC PRINCIPLE

Natural science always presupposes Man, and we must become aware of the fact that, as Bohr has expressed it, we are not only spectators but also always participants on the stage of life.

Werner Heisenberg, 1958

This remark, by one of the most famous physicists of the twentieth century, would be trivial but for the fact that it entails a peculiar view of the relationship between a behaving object and a perceiving subject. It constitutes an attack on realism that originated in quantum physics in its formative years in the 1920s and has come to have a major influence in other fields of intellectual inquiry often far removed from natural science. The principle is clearly stated by Heisenberg as follows: 'For the smallest building blocks of matter [the elementary particles] every process of observation causes a major disturbance; it turns out that we can no longer talk of the behaviour of the particle apart from the process of observation. In consequence, we are finally led to believe that the laws of nature which we formulate in quantum theory deal no longer with the particles themselves but with our knowledge of [them]' (1958: 99–100). He continues: 'The conception of the objective reality of the elementary particles has thus evaporated in a curious way ... into the transparent clarity of a mathematics that represents no longer the behaviour of the elementary particles *but rather our knowledge of this behaviour* [emphasis mine].' This is what is usually known as the Copenhagen interpretation of quantum mechanics, so called because it was popularized among physicists by Heisenberg and others who worked with Niels Bohr at the Institute for Theoretical Physics in Copenhagen more than half a century ago. It has been restated more recently by John Archibald Wheeler as follows: 'Participation is the incontrovertible new concept given by quantum mechanics: it strikes down the term "observer" of classical theory, the man who stands safely behind the thick glass wall and watches what goes on without taking part. It can't be done, quantum mechanics says. Even with the lowly electron, one must participate before we can give any meaning whatsoever to its position or momentum' (1973: 1217).

It is important to grasp what is being said here. Many acts of observation significantly disturb what is being observed – which is why psychologists and sociologists use 'one-way' mirrors (stand 'behind a thick glass wall'). Even the instruments that are used to make observations may affect what is being observed (for example, electron microscopes). But as Trigg (1980: 156) has observed, to talk of the causal effect of an observer or instrument on an object presupposes that the one can be separated from the other. Realism is preserved. The quantum physicists are saying more than this, however. They are saying that without the observer, without human consciousness, *there would be no reality*, which is precisely the position taken by idealists such as Berkeley: To be is to be perceived. That this interpretation is correct may be seen quite clearly from the following quotation from Eugene Wigner's *Symmetries and Reflections*:

> The interaction between the measuring apparatus and ... the object of the measurement results in a state in which there is a strong statistical correlation between the state of the apparatus and the state of the object ... The state of the united system, apparatus plus object, is, after the interaction, such that only one state of the object is compatible with any given state of the apparatus. Hence the state of the object can be ascertained by determining the state of the apparatus after the interaction has taken place ... *It follows that the measurement of the state of the*

object has been reduced to the measurement of the state of the apparatus. (1967: 187; emphasis added)[19]

However, Wigner continues, the measurement is not completed until 'a correlation is established between the state of the ... apparatus and something which directly affects our consciousness.' This last step, he says, 'is shrouded in mystery.' He adds: 'The very study of the external world [leads] to the conclusion that the content of the consciousness is the ultimate reality.' This, in spite of the reference to an 'external world,' is clearly a form of idealism.

It is this kind of subjectivism, not to say mysticism, that especially concerned Einstein, who remained opposed to the Copenhagen interpretation for the rest of his life and tried continually to dispel it, both in his writings and in prolonged debate with Bohr. One of the dangers of the Copenhagen interpretation that he and others (such as Karl Popper[20]) clearly foresaw was the possibility it opened for a retreat into irrationalism. In spite of Heisenberg's talk of the 'transparent clarity' of the mathematics, *this* quantum jump appeared to be a leap into the dark, with the nature of the world condemned to be forever wrapped in an impenetrable fog. Einstein did not, of course, reject the new quantum physics (which he had helped to found) as a workable explanatory and heuristic tool, one that was leading to remarkable new discoveries. No one could deny this; but he did assert that it must be incomplete, that there was something missing. Heisenberg, however – and he was followed in this by many others – asserted that it was complete; that in its essentials the theory would never be surpassed, another version of the notion that physics had reached the end of the road.

It is clear, however, that in spite of much opinion to the contrary the Copenhagen interpretation of quantum mechanics is not beyond dispute and the issue is so far unresolved.[21] Karl Popper has provided in many of his writings, and especially in his *Quantum Theory and the Schism in Physics* (1982), an alternative interpretation based on powerful arguments. Popper is a philosopher, not a physicist, though he has a considerable knowledge of physics; but there are physicists who do not agree with the Copenhagen interpretation either. The matter has been debated again as recently as 1980 by Bernard d'Espagnat, a physicist at the University of Paris-South, and Victor Weisskopf at the Massachusetts Institute of Technology. d'Espagnat's arguments were set out in an article in *Scientific American* (1979),[22] and replied to in the same journal by Weisskopf (1980). d'Espagnat confronts the often-used argument that quantum mechanics is simply a set of remarkably successful rules for computing and predicting.[23] This purely instrumental view is disputed by him on the ground that certain recent experiments have shown that either realism or QM theory must be false and that the weight of the evidence is in favour of QM theory. The physics is complex but, essentially, the experiments have to do with correlations between distant events and the principle of action at a distance. The QM interpretation of the results is that they show that these correlations *cannot be said to exist before they are measured*, thus confirming the central postulate of the Copenhagen interpretation that a quantum-mechanical system has no defined state before it is defined by the experiment; or, as Bohr would have put it, until the experimenter has decided what is to be measured and has acted on that decision, no experiment has taken place.

It is interesting to note that in a rebuttal of Weisskopf's criticisms (which led the latter to conclude that 'the ideas of quantum mechanics do not contain any reasons whatsoever for giving up the concept of a reality that is independent of mind'),[24] d'Espagnat argues (1980) that quantum mechanics 'can [yet] form the basis for an objectivity of a new kind: a man-centered objectivity that may well suffice for the building up of the entire body of positivist science.' It is clear from the context in which this remark is made that the 'positivism' envisaged must embody a sensationalist-empiricist epistemology of the kind advanced by Mach: that is to say, one that reduces knowledge to the mind's interpretations of what 'appears' to the senses – in this case, measurements. As we have seen, such an epistemology runs so contrary to the actual experience of science and scientific discovery that it cannot possibly be regarded as satisfactory.[25] The notion of a 'man-centered' objectivity is, moreover, entirely consistent with Bohr's famous dictum that 'It is wrong to think that the task of physics is to find out how nature *is*. Physics concerns what we can say about nature.' But as Trigg notes, 'This repudiation of ontology undermines the whole purpose of physics. What is the point of saying anything about nature, if we are not attempting to say how nature is?' (1980: 172). The obsession now so prevalent in philosophy,[26] and, by extension, in the philosophy of science, with 'what we can say' – the idea that the limits of our language define the limits of what we can know (since knowledge can only be expressed in language) and that we are forever trapped in this framework – is surely absurd since it leaves *learning* unexplained; it cannot account for the growth of knowledge.

Many scientists today would agree that life and consciousness, whether confined to our own small planet or, as some argue, in view of the vastness of the universe, certain to exist in many other places,[27] are mere epiphenomena – a kind of cosmic accident explainable by the fact that local conditions just happened to be right for their emergence. There are now, however, a number of prominent cosmologists, including John Archibald Wheeler, who argue precisely the opposite: that the *requirement* that life and consciousness 'should exist somewhere, for some time,' is built into the structure of the universe (Wheeler 1975: 576). This is the so-called anthropic principle.[28]

The generally accepted cosmological theory, which is now supported by much evidence, is that the universe had a beginning in a 'Big Bang.' Starting from a dimensionless point of infinite density called a 'singularity,' the subsequent 'explosion' created space and time, all the constituents necessary to produce all the matter and energy that now exists in the universe, and all the physical laws that govern it. On the basis of the theory, and empirical evidence about its present state, the emerging structure of the universe during the first three minutes of its life has now been calculated in some detail. The normal mode of scientific reasoning is to predict an event or the state of a system on the basis of a theory and a set of initial conditions, from the conjunction of which the prediction, or outcome, can be logically deduced:

Premisses	Theory (which may be itself, or depend upon, a physical law or laws) Initial conditions
Conclusion	Predicted event or state

This procedure is not possible in this instance. The cosmologists know the outcome, the present state of the universe, and they possess a (well-corroborated) theory. What they do not know, and have to retrodict, are the initial conditions that brought about the present state. They are also obliged to assume that the laws of physics that operated in the early moments of the creation have not changed. This assumption provides the logical basis for the anthropic principle. The empirical argument is that the universe appears to be governed by a few basic laws (perhaps no more than fifteen) that, had they been slightly different, would have produced a very different universe from the one we inhabit – in fact one in which life and consciousness would not have evolved. For example, had the force that governs the atomic nucleus been only slightly weaker than it is, the only stable element would have been hydrogen, and, as the physicist Robert Dicke wittily observed, 'it is well known that carbon is necessary to make physicists.' Had the force of gravity been slightly weaker than it is, matter would not have 'congealed' into stars and galaxies. If it had been somewhat stronger than it is the sun (or any other average star) would not have survived long enough for a planet to exist that was capable of supporting life.[29] Hence Wheeler's question: Has the universe had to adapt itself from its earliest days to the future requirements of life and mind?

Stephen Hawking is quoted by John Boslough (see note 29) as saying that according to one version of the anthropic principle there are a large number of possible universes, but in only a small number of these will there be the same conditions and parameters as in our universe: 'In those it will be possible for intelligent life to develop and ask the question "Why is the universe as we observe it?" The only answer will be that, if it were otherwise, there would be nobody to ask the question.' As Boslough remarks, 'This rather curious concept' (the anthropic principle) is decried by some scientists on the ground that it offers no explanation at all; and it is not difficult to see why they should say this, for there does seem to be something circular about it. In any event, there would appear to be nothing in the principle as such that is inconsistent with realism.[30] In combination with the orthodox version of the 'meaning' of the quantum revolution, however, this principle reinforces the idea that we are participants in the universe in a special sense: that if we were not here it would not exist. This is the line of thought followed by Professor Wheeler. He argues that the quantum principle must be considered to be the overarching feature of nature because of its astonishing productiveness and its fertility in explaining so many features of the world – 'all of chemistry and most of physics,' it has been said, to which may be added 'and much of biology also.'[31] It follows (for Wheeler) that since the participation of the observer is inescapable at the subatomic level, it must be so, given the omnipresence of the quantum principle, at the level of the cosmos also. 'No physics without an observer' applies universally; the cosmos itself is, 'in some not yet clearly understood sense,' a product of the consciousness that it has produced; all the physical laws on which the universe depends themselves depend on the existence of a knower. We are back once more, whether Professor Wheeler intends it or not, with Berkeley: there can be no reality without an observing subject or, as Wheeler more flippantly puts it: 'What is the good of a universe if there is nobody around to look at it?'

Few scientists, as Gale has pointed out (1981: 171), are prepared to follow Wheeler into this mysterious realm. He 'has carried the anthropic principle far beyond the domain of the logic of explanation; he has crossed the threshold of metaphysics.' It may be added that he has done so without considering that the orthodox interpretation of quantum mechanics may be wrong.[32] As Einstein said, the most incomprehensible thing about the universe is that it is comprehensible. It is made less, not more, comprehensible by an unnecessary retreat into idealism. But perhaps even this is to be preferred to the sad descent of some physicists into Oriental mysticism. I do not intend by this remark to decry Oriental mysticism – in its place – but, in spite of the attempts by some recent writers, such as Fritz Capra,[33] to convince us of the contrary, it is utterly inconsistent with science.

SCIENCE AND OBJECTIVITY

Wherever universal agreement prevails, the realist will not be the one to disturb the general belief ... For according to him it is a consensus ... that constitutes reality.

Charles Sanders Peirce

The American philosopher C.S. Peirce (1839–1914) was a realist of sorts, but he held that, since realism cannot be proved, our knowledge of the world can be nothing but belief. More interestingly, he argued that what is 'true' and 'real' in science is established by a consensus of all those who are competent to make a judgment – the community of scientists. In this he anticipated the ideas of a large and currently influential group of sociologists and philosophers of science.[34]

The sociology of knowledge has traditionally been concerned with the problem of the *generation* of knowledge: with the social processes by which knowledge is created and the social factors that shape its content. Modern sociologists (and philosophers and historians of science who are influenced by them) operate generally within this tradition. But they oppose their predecessors, such as Karl Mannheim, who exempted science from the otherwise universal principle – that knowledge is socially constructed and situationally conditioned – because of its objective character, though their reasons for holding it to be objective were rarely clearly stated.[35] Modern interpreters reject this privileged status for scientific knowledge. It is, they say, in respect of its social construction, no different from 'ordinary,' or common-sense knowledge. Their studies of the processes by which working scientists generate their knowledge claims have led them to the conclusion that scientific knowledge is nothing more than a system of provisional beliefs about the world arrived at by negotiated consensus. They overlook the fact that the provisional acceptance of a proposition about the world (a truth claim) is not the same thing as a belief in its truth. Beliefs inevitably lead to demands for justification. A statement such as 'scientific knowledge is nothing but justified belief' leads to the question: 'But how is this belief justified?' This in turn leads to attempts to find warrants for believing: in the evidence of the senses, in intuition, in lack of doubt, in consensus (as in the present case), or in some other base. All these grounds for believing are *subjective*, either to an individual or to a collectivity. But the search for justification – for warrants for belief – is the search for a refuge from unbelief and is thus the antithesis of critical inquiry. If science is directed towards the increase of knowledge, towards greater understanding, then it must rest on the practice of criticism, not on justification. Indeed progress often requires a suspension of belief.

John Ziman has written: 'It has been put to me that one should ... distinguish carefully between Science as ... knowledge and Science as what scientists do ... This is precisely the sort of distinction that one must *not* make' (1968: 11; emphasis his). On the contrary, it is precisely a distinction that *must* be made. It is true that in the practice of science we find consensus and conventional agreements on such matters as the manner of determining the reliability of evidence, the assessment of professional competence, whether or not one experiment actually replicates another as claimed, the reliability of instrumentation and the consistency of a theory with other related theories, and even (as we shall see later in chapter 5) as to whether certain ideas are or are not to be regarded as scientific. There are a host of such technical and cognitive questions that are subject to debate and provisional resolution. But it is a serious mistake to conclude, as does Mulkay (1970),[36] that the knowledge claims advanced by scientists are themselves *nothing but* claims 'which have been *deemed to be adequate* by particular groups of actors in specific social and cultural contexts' (emphasis added). The mistake lies precisely in failing to distinguish between what scientists do and what they produce. The practice of science is a process of producing scientific knowledge, but the sociologists' assumption that once the production process has been laid bare there is nothing more to be understood is false.

Knowledge is an artefact; it is created by us; but once created it exists outside ourselves; it possesses a certain autonomy; it affects us (it has unintended consequences), and we can affect it, for example by criticizing it, by examining its logical structure; it is *real* (see above, p. 7). It is, in a word, *objective*. The sociology of science looks at the behaviour of scientists producing knowledge, at the production process; it is concerned with how scientists get their knowledge, with its social causes. But this is only half the story, and totally misleading in the absence of the other half. Whence, and how, knowledge arises are not unimportant matters; but of much greater importance is the *objective character* of knowledge claims once they are created.

It is, in fact, necessary to make a sharp distinction between the *process* of arriving at knowledge and the *content* of what is claimed to be known. This can be illustrated by thinking for a moment about the difference between two types of statements. Some statements may be to the effect that an idea or a problem is 'muddled,' 'puzzling,' 'interesting,' and so forth. These statements incorporate *subjective* terms. The terms relate to mental states. Other statements, about ideas or propositions, may be to the effect that they are 'contradictory,' 'incompatible,' or 'inter-deducible.' These statements incorporate *objective* terms, terms that have nothing to do with mental states. A different kind of example is one suggested by Michael Polanyi. Consider a map. If the map is wrong and misleads me, I blame the map and not myself. But if the map is correct and I fail to read it correctly and am thereby led astray, I blame myself and not the map. The map is objective in both these ways. It is also objective in a third way, which Polanyi did not mention: the map contains information that I may not notice, *but others may*. The same can be said of scientific theories. They are products of human minds, but once produced they are as objective as the map. A theory may have consequences that are not recognized at the time it is proposed – Einstein's theory of general relativity, for example; and it may not be fully understood even by those who propose it. It is said, for example, that Einstein did not fully understand his own principle of covariance. As Popper has said: 'Knowledge in the objective sense is totally independent of anybody's claim to know; it is also independent of anybody's belief, or disposition to assent, or to assert, or to act' (1972: 109). This view is consistent with the argument made earlier in this chapter, that science (indeed all empirical knowledge) begins, not with observations (which are subjective), but with problems (which are objective – they *are* 'given'). The same is true of the solutions to problems. While the process of *seeking* solutions is subjective (at least in part) and involves states of mind – feelings of frustration, for example – the solution itself is objective.

One of Popper's great contributions to the philosophy of science is his deceptively simple idea of a 'world' of objective knowledge. He distinguishes this 'world' both from the 'world' of physical objects and processes, whether naturally existing (for example, atoms, the replication of living cells) or created by us (bridges, the recombination of genetic material in a laboratory), *and* from the 'world' of conscious experiences (thinking, feeling hungry, remembering, believing, and so on). These worlds are to a high degree autonomous, but they are also interconnected in that they affect (have a feedback effect on) each other. World 1 (the physical world) and World 2 (the world of conscious experiences) can and do interact directly, as can and do Worlds 2 and 3 (the world of objective knowledge); Worlds 1 and 3 can interact only through the mediation of World 2. A book is a physical object (World 1), but it also contains ideas (World 3), and it is only through mental states (World 2) that those ideas (which may be ideas about World 1) can be grasped by us. These ideas may be correctly grasped by you but incorrectly grasped by me, and this is possible because they exist independently of both our minds. In the same way, a scientific theory is an 'object' in World 3, perhaps about World 1. It is tested by reference to World 1, but this action invokes processes occurring in World 2. Knowledge (and particularly scientific knowledge) is only in one sense 'a state of mind,' and only in one sense a product of the social and historical situation in which it was produced. In another more important sense it exists (particularly when it is recorded publicly in recoverable form) independently of the minds that produced it, *or even of any minds at all*, and outside of particular time and place. The knowledge that was locked in the (objective content of) Linear B script was not known to any minds for centuries, but since the deciphering of the script, it is now known – to us. We can even say that knowledge may exist before it is discovered. Numbers were invented by humans, but the knowledge that there are, for example, prime numbers and irrational numbers, was a discovery. This knowledge was not 'socially constructed.'

Most important, perhaps, this conception of objective knowledge establishes a definitive difference between idealism and realism. Idealism holds that reality is created *by* us and exists *for* us alone (it is our world and 'reality' is 'reality-for-us'). Realism holds that reality is not solely our creation and is not simply 'our' reality because, although we may help to create it, that is add to it, it poses problems for us and surprises us (it adds to us), it grows independently in an important sense of us, and it transcends us here and now because it will also pose problems for our successors and surprise them too.

2

Science and Technology

Science is the transformation of Nature into concepts for the purpose of mastering Nature; it belongs under the rubric *means*.
Nietzsche, *The Will to Power*

Nietzsche's aphorism is nowadays widely believed. Expressed less harshly, many – perhaps most – people today believe that science exists primarily for its instrumental application as technology, and that the two forms of human activity are essentially one; that they have become, in the words of Jurgen Habermas, 'fused in a single system' (1970: 104). One of the evident consequences of this belief has been the assumption that science must be treated as an 'investment good'; that it is to be valued solely for its contribution to the material benefit of society, justified and supported only in so far as it promises a clearly foreseeable 'pay-off' in terms of practical application for economic development, national security, social welfare, and other 'national goals.' This has been the assumption behind most of the pronouncements by national governments and international agencies about 'science policy' in the past twenty-five years. These aims are not unworthy. The 'improvement of Man's Estate' (in Francis Bacon's words) has long been regarded as a proper objective for scientific inquiry. But a purely instrumental view of science obscures its value as an embodiment of some of the highest expressions of the human spirit. An extreme, though not untypical, version of the instrumental view of science is expressed in the following passage by a political theorist (Eisenstein 1977): 'Science, as all thought, is a process of practical activity. Its justification is that it works to resolve practical problems. The ultimate justification and explanation of science is not truth or correctness; rather it is engineering. If it was not for engineering no one would indulge in science.'

I believe all such views of science to be seriously mistaken. The analytical and practical identification of science with technology falsifies history, lacks empirical support, and rests on a profound misconception of the logic of the two modes of knowledge and practice. Science and technology are two distinct, *though interacting*, complexes of knowledge and practice, persons and institutions, beliefs and values. The economist Theodore Schultz, writing about science as an 'economic good,' has said that you always get wool and mutton when you produce sheep (1980). My answer to this is that you do not get both from the same sheep (at least if you want good mutton and fine wool). The interesting question (to continue the metaphor) is how the two systems of mutton producing and wool producing interact. One of the objectives of this chapter is to suggest at least a partial answer to this question.

THE HISTORICAL RELATIONSHIP BETWEEN SCIENCE AND TECHNOLOGY[1]

Until about the seventeenth century in the Western world (the situation in China was quite different)[2] science and technology went their separate ways. Technology was for centuries created in the course of the exercise of crafts that used relatively simple techniques based on knowledge acquired through generations of trial and error and transmitted by imitation and apprenticeship. The entire early history of Western technology (and, indeed, of Oriental technology too) gives the lie to Francis Bacon's dictum that 'where the cause is not known the effect cannot be produced.'

Where science and technology touched it was usually science that benefited. For example, the replacement around the fourteenth century of Aristotle's theory of motion, based on the principle of push, by the theory of impetus (an idea put forward by Jean Buridan, among others) was probably suggested by the example of the rotary grindstone, when it was observed that the stone continued to rotate after the grinder's hand was removed from the crank. A second example is the weight-driven public clock, which seems to have made its appearance in Europe sometime in the latter part of the thirteenth century and which not only inspired the cosmological image of a mechanical universe (Cardwell 1973), but also probably led to the eventual acceptance of the notion of absolute time.[3] It is also suggested, perhaps with less veracity, that Sir William

Harvey may have got the idea of the action of the heart as a pump from the example of the bellows used to raise water.[4]

In the next phase, coinciding with the first industrial revolution in Britain – roughly from 1750 to 1840 – a movement began from craft-based technology to science-based technology, from technology based on knowing how to technology based on knowing why. It was the beginning of a technology that expressed concern for understanding the scientific principles on which technological progress could be founded. Self-taught artisans began to be replaced by innovating entrepreneurs who actively sought the assistance of men of science. An illustration of this change is to be found in the organization, around 1768, of the Lunar Society of Birmingham, so called because its members met for discussion on Mondays nearest to the full moon. The members included men like the natural philosopher William Small (one of whose pupils had been Thomas Jefferson), the metal manufacturer Matthew Boulton, Boulton's partner James Watt, Erasmus Darwin, and Joseph Priestley. Benjamin Franklin was a corresponding member. R.L. Edgeworth, a wealthy land-owner, who was also a member, provides an interesting glimpse of the way such associations were perceived. He wrote: 'The knowledge of each member ... becomes in time disseminated among the whole body, and a certain esprit de corps, uncontaminated by jealousy, in some degree combines the talents of the members to forward the views of a single person.'

The Lunar Society helped to inspire the formation of many other 'literary and philosophical societies' in the late eighteenth and early nine-teenth centuries. One of the first and most famous of these was the Manchester 'Lit. and Phil.' which was founded in 1781 and survives to this day. Others soon followed in the industrial centres of the North of England and the Midlands. They were essentially meeting places for local merchants, industrialists, and men with a scientific or inventive bent. Their purpose was the dissemination of scientific and technological knowledge rather than its advancement, and the spreading of the new scientific-technological culture. In 1785 the president of the Manchester Society stated: 'We have established a society which, in its views, combines practice with speculation; and unites, with the culture of science, the improvement of the [mechanical] arts.' At the national level, the Royal Society for the Encouragement of Arts, Manufactures and Commerce was established in 1754. Characteristically, it originated in a London coffee-house, and its founding fathers were a group of noblemen, clergy, gentlemen, and merchants. A further harbinger of the new age was the formation in London in 1799 of the Royal Institution, the stated purpose of which was 'to diffuse the knowledge and facilitate the general introduction of useful mechanical inventions and improvements,' and to teach 'by courses of Philosophic Lectures and Experiments the application of science to the common purpose of life.'

The application of scientific research and the scientific attitude to technological practice was, however, a slow process, a process that was by no means complete even by the end of the nineteenth century. Many important developments were necessary for its completion. The social and organizational structures of science and engineering had to undergo change; systems of technological and scientific education and training had to be created, and existing institutions, especially the ancient universities, had to be persuaded that scientific research, and the teaching of science and its applications, were proper activities for institutions of higher learning; changes were necessary (at least in Europe) in the attitudes of governments to the support and sponsorship of science and technological research and development; and, lastly, changes were required in the outlook of industrialists themselves.

Many of the social and organizational features of science and engineering that we recognize today were developed during the course of the nineteenth century: the emergence of the 'scientist'[5] and the civilian 'engineer' as distinctive, and distinct, social roles freed from the taint of amateurism; the recognition of science and engineering as providing lifetime careers, entry to which had to be obtained by demonstrated competence certified by examinations and the awarding of diplomas or degrees; and the organization of scientists and engineers in specialized professional bodies, often publishing their own journals. Germany took the lead in Europe in moving science into the universities; and especially, in the *technische hochschulen*, created outside the university system, in melding science with technological education. The education of a technological elite had begun in France with the creation of the *grandes écoles* during the second half of the eighteenth century, but one of the major contributions of the *technische hochschulen* was to train for the middle ranks – the technicians of industrializing Germany. Nor were they merely teaching institutions; they also maintained effective research laboratories. The research laboratory was essentially a nineteenth-century invention, one of its prototypes being Justus Liebig's famous chemical laboratory, which he established at the University of Giessen in 1825. The resistance of the established universities to the new scientific and technological culture, though not absent in Germany, was much greater in Britain, and it led to the creation of new universities and university colleges in London and several of the provincial industrial cities where there was freedom to establish chairs in the applied sciences and engineering.

It was otherwise in the United States.[6] For example, the Lawrence Scientific School at Harvard was founded in 1847 specifically for those who intended 'to enter upon an active life as engineers or chemists, or, in general, as men of science applying their attainments to practical purposes.' At Yale a School of Applied Chemistry was established about the same time, merging with the university's school of Engineering in 1851. The Land Grant colleges, which were established under an act of the United States Congress passed in 1862, had specifically utilitarian aims: the promotion of agriculture and engineering. The Massachusetts Institute of Technology was established in 1861, on the eve of the Civil War, and many other similar institutions followed in later years including the now equally famous California Institute of Technology (originally the Throop Polytechnic Institute). One of the consequences of the emphasis on utility, however, was the neglect of large areas of pure science. This, by the third quarter of the century, was causing serious concern to many American scientists. One physicist, addressing the American Association for the Advancement of Science in 1883, lamented: 'We are tired of seeing our professors degrading their chairs by the pursuit of applied science instead of pure science,' adding with a certain bathos, 'lingering by the wayside while the problem of the Universe remains unsolved.'

Meagre state support for science and its application to technology was a universal problem in Europe, at least outside Germany, for most of the nineteenth century; and this in spite of continual complaint from the scientific communities. English scientists complained of how much better off were their colleagues in France, and both complained to their governments about the disastrous neglect of science by comparison with the situation in Germany. It was German scientific pre-eminence, and especially German industrial and technological pre-eminence, that eventually persuaded politicians of the important role that science could play in developing technology.

The nineteenth century witnessed the beginning of the twin processes of 'scientizing' technology and establishing it as a fully systematized corpus of knowledge and practice. Modern technology is quite different from pre-modern technology in both these respects: it is science-based and it is systematic (Hannay and McGinn 1980). Yet, although it is science-based and tends to be cultivated in somewhat the same manner as science, systematic technology also possesses an internal dynamic and an internally generated body of knowledge that is quite unrelated to scientific discovery (Polanyi 1958: 178–9). One of the myths that has grown up in the past forty years, and which, for a while at least, coloured thinking about governmentally directed science policy, was that technical innovations are the direct result of advances in science. This myth has now been largely dispelled as a result of intensive study by economists and other students of technological innovation. Schmookler (1966), for example, found that in many of the nine hundred inventions he investigated there was no identifiable scientific component at all, and that in almost every instance the stimulus that produced the invention was economic, not scientific. In another study (Langrish et al. 1972), eighty technological innovations were examined, and the conclusion was that all but a few were based on previous technology.[7]

There is, moreover, little reason to doubt de Solla Price's thesis (1965) that, where technology arises from science, it is more likely to be from the established trunk and main branches of the tree of science than from its growing twigs. It is 'old' science, 'packed down' science (in Price's graphic phrase), science that has passed into the textbooks, or at least 'thoroughly understood and carefully taught science,' that technology uses most. Or, to borrow another formalization, this time from Wolf Schafer and his associates (1983), technology uses scientific theories that have entered upon their 'post-paradigmatic,' or 'finalised' stage; that is to say, those parts of scientific disciplines that have become sufficiently mature for them to be looked to for their practical use. Price's conclusion was that the interaction of scientific and technological ideas takes place, not through the current research-front literature, but rather through education and training, chiefly of course in the universities. The science that technologists absorb, he suggests, is an already digested science, and research-front technology is strongly related only to that part of scientific knowledge that has become part of the accepted canon. It may be that some modern technologies, such as genetic engineering, constitute exceptions to Price's thesis; but they are likely to remain the exceptions rather than the rule. Price argues (1965) that the naive picture of technology as applied science simply will not fit all the facts. From time to time certain advances may derive from the injection of new science, but in the main old technology breeds new. Science and technology today behave like a pair of dancers to the same music, but they hold each other at arm's length instead of dancing cheek to cheek.

THE INTERACTION BETWEEN SCIENCE AND TECHNOLOGY

We shall now examine briefly some of the steps in this dance.

1 / Although the idea that there is an unbroken continuum from scientific discovery through applied research and development to technological application is almost certainly false, there are nevertheless clear cases where breakthroughs in pure science do form the basis for the development of new technologies. The discovery of artificial radioactivity by Irene Curie and her husband Frederic Joliot in 1934 is one example. This was a purely scientific achievement. Indeed it came about when the discoverers were investigating something quite different,[8] namely the emission of positrons and electrons from light elements when bombarded by alpha particles; yet it led to the founding of a technology for producing artificial radioisotopes that could be used as tracers[9] in the human body; in biological, geological, and archaeological research; in industry; and as sources of radiation in the treatment of cancers. It was also one of a series of crucial discoveries in the chain that led to the exploitation of nuclear energy. A contemporary example that parallels, and may well surpass, this discovery in terms of its technological implications is genetic engineering, which has been made possible by advances in microbiology, especially since the discovery of the structure of the DNA molecule by Crick and Watson in 1953. The pure science that followed from this discovery required the invention of techniques for altering the genetic make-up of living cells. These techniques were developed for research purposes, but they also had practical uses. They constitute an example of the movement from science to technology (in this case, industrial, agricultural, and medical technology) through technology developed originally for purely scientific ends, a phenomenon that we shall look at further later in this section.

2 / It is evident that much science is done today in support of government-run or governmentally inspired 'missions,' that is, specifically to advance the end of technological application. The classic, though by no means the only, examples are military or paramilitary: nuclear weapons, much laser and microwave research, research into chemical and biological warfare, and the American and Soviet space programs. The prototype 'big' mission was the code-named 'Manhattan District' of the Second World War. The mission was unequivocal: to produce an atomic bomb, and to produce it in the shortest possible time at whatever the cost. There was then no 'nuclear' technology. The Manhattan District project brought together many of the world's leading scientists and en-

gineers to create it. In the event, industrial plants were already under construction before many of the engineering problems had been solved, and applied research was being done before all the problems in theoretical physics and chemistry had been worked out. All the supposed normal sequences of research and development were broken in the race against time, and scientists who had hitherto been concerned solely with advancing knowledge found themselves caught up in the exciting challenge of applying their discoveries to a technological end. At the wartime Los Alamos Laboratory, where the uranium and plutonium bombs were invented, the organization of work and the approach to research took on many of the characteristics of the world's larger industrial laboratories, such as those of the Bell Telephone Company. The scale of research was big; output was the product of teams rather than individuals; above all, the pure science that was done became 'fundamental' or 'basic' to the achievement of a specific end.

Today, many of the largest corporations operate laboratories in which pure science is done alongside applied. Bell Telephone was one of the pioneers. Today, the company employs in its research laboratories more than 19,000 people of whom more than 2500 possess doctorates.[10] I referred at the beginning of this chapter to science and technology as two distinct, though interacting, complexes of knowledge and practice, persons and institutions, beliefs and values. It is in the second of these groupings that the two forms are least sharply differentiated. Corporate research provides two important illustrations of this aspect of the science-technology relationship. The first is that an individual may be a scientist at one point in his or her career and a technologist at another; though this does not mean that the distinction between science and technology is purely arbitrary, as some (such as Mayr 1976) have tried to argue. The second is that the location of science – the place where it is done – tells us nothing. The science that is done in universities is not all pure science, and the science that is done in industry is not necessarily all applied, or technological science. It is more pertinent to ask: 'Why is this research being done?' and '*For whom* is it being done?' Bell was a pioneer in setting some of its qualified staff to work on purely scientific problems in the hope that the solutions might lead to practical use, and that some at least of the solutions might be obtained by Bell scientists before they were reached by university scientists. For that reason the scientists employed by Bell in pure research have long been given the same freedom to work and the same kind of research environment that is enjoyed by university scientists (Hoddeson 1980). Nevertheless, when pure science is done in industry it is done in the hope of future profit or for reasons of institutional prestige. It is worth noting that several scientists employed in industrial laboratories have won Nobel Prizes; for example, C.J. Davisson in 1937 (physics) and William Shockley in 1956 (also physics), both Bell employees,[11] and Geoffrey Newbold in 1979 (physiology), an employee of Electrical and Musical Industries Ltd. Davisson's discovery of electron diffraction by crystals provided experimental confirmation of one of the theoretical implications of quantum mechanics proposed by de Broglie; Schockley 'discovered' the transistor; Newbold was awarded the prize, jointly with Alan Cormack, a university scientist, for their work (independently of each other) on the development of computed tomography – an X-ray imaging technique.

3 / Although it is often possible to forecast the general directions in which scientific discoveries will be made, because work on the frontiers of knowledge is usually concentrated in those areas where current research is scientifically the most fruitful (that is to say, is producing the most interesting results), specific applications of these discoveries are not always so easy to predict. The reason is that many of the most important advances in theory and experiment in science have implications of which the original discoverers, and their peers, are unaware at the time. Such is true of the laser, the principle of which was embodied in a purely theoretical paper published by Einstein[12] in 1917. In fact, reality in this case was a little more complex, since Alexander Graham Bell already had an application in the use of a beam of light to carry telephone calls – an idea that he patented some seventy-five years before the basic principles for the construction of a laser were finally worked out by two scientists in the laboratories that bear his name.[13]

The complexity of the relationship between pure science and technology is perhaps nowhere better illustrated than in the story of the transistor, another of the great socially transforming innovations of the late twentieth century. This story has been explored in detail by Gibbons and Johnson in *Nature* (1970). In brief, the invention of the transistor was preceded by a century-long history of pure research into the phenomenon of semiconduction, dating back to a discovery made by Michael Faraday in 1833. This theoretical and experimental work culminated in some papers on the quantum theory of semiconductors published by A.H. Wilson in 1932 – 'culminated' in the sense that Wilson's theory was the first to provide 'the much needed model of conduction and rectification phenomena,' that is, the pure science that could explain, among many other things, how a transistor might be possible. But it would be presuming too much, Gibbons and Johnson argue, to deduce from this that Wilson's model led directly to a piece of technology. On the contrary, they argue: 'Although Wilson's principal contribution was to provide a theory of conduction in semi-conductors, it is apparent from the earliest history of semi-conduction that research was conditioned both by the development of associated technologies and by commercial and social stimuli, some of them military.' One of these associated technologies was radio, begun in the early 1900s; another was radar, developed for military purposes during the Second World War. Shockley's 'invention' of the transistor was, as he said himself in his Nobel lecture, the result of his strong interest in the practical applications of solid state physics. Gibbons and Johnson conclude that, although Wilson's model was of great importance, the early history of the transistor suggests that it 'arose primarily as a result of technology building on technology – in this case rectifier technology.' Their case study supports de Solla Price's view that the relationship between science and technology is symbiotic – a biological term that has been defined rather nicely as 'a state of affairs in which two often very dissimilar organisms live together in mutual dependence and for mutual benefit, each making up for the other's shortcomings.'[14]

4 / Science makes use of technology and may make new demands upon it. In principle, there is nothing new about this. The development of the two paramount 'aids to the senses' (Bacon's phrase) in early modern science, the telescope and the microscope, rested on improvements in glass manufacture. Galileo found it necessary to grind his own lenses for his telescopes and, in the absence of an accurate chronometer, to adopt various ad hoc devices such as singing in order to measure the lapse of time. People who argue that the experimental and observational data collected by the ancients were often 'fudged' forget how simple and inaccurate were the available instruments and how greatly experimenters and observers were forced to rely on qualitative judgments.[15] In an address to the Lycée Pasteur half a century ago, the great French physicist Louis de Broglie said: 'Every important step forward made by astronomy, physics, chemistry or biology had one essential condition – the previous existence or invention of certain apparatus; and as the sciences sought to extend their advance, so it became necessary for experimental technique to develop and to expand in its delicate adjustments ... Left to itself, theoretical science would always tend to rest on its laurels.'[16] But we should note that de Broglie stressed improved experimental technique as well as better apparatus. Ptolemy's calculations of the refraction of light were certainly 'fudged' (this can be seen from

inspection of his results, which are recorded); but this seems to have come about less because of the crudity of his apparatus than because he had no tradition of experimental technique to guide him. He simply thought that a certain sequence of numbers 'looked right.' As has often been said, the legacy that the great medieval alchemists left for the new discipline of chemistry in the seventeenth century was a tradition of careful experimental technique. Nevertheless, experimental techniques could hardly progress in the absence of improved instrumentation.[17] This fact is strikingly illustrated in a passage from the journal *Scientific American*, August 1884:

> The brilliant discoveries by Pasteur and by Koch are as much due to the perfected microscope as to any one cause. The nature and habits of the tubercular bacillus have only been capable of study since the microscope was so improved that organisms heretofore unrecognisable stand revealed ... And when another potent servant of man, electricity, is summoned to aid the microscope, the power of the latter is increased to an astonishing degree. Recently in London such an apparatus threw upon a screen the image of a cholera germ, magnified two million times, and in which these long hidden and minute organisms appeared the size of a human hand.

Until the early years of the present century, however, experimental laboratories remained relatively primitive and small by modern standards. The construction of the world's first practical cyclotron for use by Ernest Lawrence at the University of California in 1932 effectively ushered in the era of Big Science, to coexist with 'small' science. Much of today's science requires the construction of, for example, large radiotelescopes, plasma generators, and particle accelerators for high-energy physics. These are not simply 'instruments'; they are research laboratories developed *around* instruments; and they require for their operation a great deal of supporting technology, such as sophisticated computers and cameras. In the initial stages, at least, of the development of this 'hardware' the engineering may itself by highly innovative. The guidance system of the Large Space Telescope that NASA proposes to put into orbit (using a space shuttle) is said to require a capacity equivalent to locking on to a single hair at a distance of two miles. If the launch is successful and the whole assembly works, the telescope will be able to 'see' seven times farther than the largest existing terrestrial telescopes, that is to say, almost to what are currently thought to be the limits of the universe.

Unlike an X-ray machine used for detecting flaws in metals or an oil-cracking plant, radio-telescopes, particle accelerators, and the like are much more than mere technological devices; each is part of a system for making scientific observations that are required by complex scientific theories. Hence the technology that is required to make them is not simply empirical engineering; it is engineering that must conform to the requirements of those theories. A piece of empirical engineering like a watermill can be said to embody in its construction certain theories about the natural world, but it has a purely practical purpose. The construction of a radio-telescope must also conform to the laws of mechanics if the telescope is to function; but the finished product embodies not only these laws but also the theories of astrophysics for the further development of which the instrument was required in the first place.

5 / The laser provides an excellent example of another type of connection between science and technology, namely, the movement from science to technology through technology developed for scientific purposes. The laser was initially a piece of instrumentation for research, and, indeed, research continues to constitute a substantial segment of the market for lasers. The technology is now used, or is being developed, for a wide range of other purposes, however; in industry for welding, in medicine for laser surgery, in defence and communications, and in work on fusion nuclear reactors, for example. In surgery its use is potentially revolutionary, since it replaces the traditional need for the removal of organs from the body by repairing them in place. The great advantage of laser surgery is that it is so precise and delicate that it causes minimal bleeding, scarring, and pain, and it does so little damage to surrounding tissues that conditions previously thought to be inoperable can now be corrected. One of the more dramatic uses of lasers in research and, potentially, again in medicine is in genetic engineering, since the laser can be adapted to produce a point of intense coherent light so small that it can remove specific genes from the chromosomes of a single cell.

The laser represents only one of the latest instances of the commercialization of technologies that were originally developed for purely scientific research. Earlier examples include tools and techniques such as high pressures, high vacuum, spectroscopy, vapour-phase chromatography, and the cathode-ray tube, which, originating in the physics laboratory, became the basis of the modern television picture tube.[18]

6 / Lastly, technology sometimes leads directly to scientific discoveries, or promotes the process of scientific discovery. This can come about in two ways. Discoveries may arise in the course of proposing scientific

solutions to technical problems; and advances in technology can lead consequentially to advances in science. A classic case of the former in the history of science is that of Galileo and the suction pump. The question why a suction pump is incapable of lifting water more than about thirty feet was of no concern to the miners of Galileo's time, or to the peasants who wished to irrigate their terraced hillsides, for they simply used relays of pumps. It did trouble Galileo, however, particularly since the Aristotelian doctrine that nature abhors a vacuum was still widely accepted. The correct response to this technological problem was not that the pumps were faulty but that there was something wrong with the theory. And this led to a new theory, that air has weight, the theory of atmospheric pressure; to a new piece of technology, the barometer, invented by Galileo's pupil Torricelli; and later still, to another fundamental principle of pure science, Boyle's Law. It was precisely discoveries like the scientific explanation of the suction pump that led in time to a growing confidence in the promise of a new technology based on science, and to the progressive abandonment of technologies based solely on common sense and rule of thumb. Unfortunately, they also led many persons to the mistaken belief that traditional technologies gave rise to modern science. One such person was the influential philosopher John Dewey, who gave wide currency to the idea. He wrote: 'Technologies of this kind [i.e., simple crafts] give that commonsense knowledge of nature out of which science takes its origin. They provide not merely a collection of positive facts, but they give expertness in dealing with materials and tools, and promote the development of the experimental habit of mind' (1948: 12).[19]

This characteristically pragmatist view of science, that it is based on 'facts' and is merely a more sophisticated common sense, is quite mistaken. All the major advances that gave the Scientific Revolution its unique character – the triumph of heliocentrism, the overthrow of Aristotelian ideas about motion, and so on – involved great conjectural leaps of the imagination, from the contemplation of everyday things and processes to the imagining of abstract possibilities. Moreover, what, in this context, is 'common sense'? What is common sense for us today was far from common sense even to the greatest intellects of the fifteenth century. It was said around 1920 that there were only two people in the world besides Einstein who understood the theory of general relativity (Sir Arthur Eddington is reputed to have asked 'And who is the other?'), but today it is familiar fare for any undergraduate majoring in physics. And the idea that science is a craft that requires only the mastery of the proper techniques is part of the myth propagated by Francis Bacon. His famous methodology – or recipe for doing science – was, he said, such that 'leaves but little to the acuteness and strength of wits, and places all wits and understandings nearly on a level' (Farrington 1973: 117).

The classic case of a technological innovation that gave rise to an advance in scientific knowledge (the second of the ways in which technology may precede science) is Marconi's invention of the wireless, which led to the discovery of the ionosphere. More than a century earlier the accepted theory of heat, the 'caloric' theory, was first challenged by Count Rumford[20] as a result of observations he made while engaged in the business of boring the barrels of brass cannon in a military workshop in Munich. His experiments led him to propose an alternative theory, that 'heat is motion' and not the massless, imponderable substance added to matter that the caloric theory required. Sadi Carnot, forty-three years Rumford's junior, whose work *Reflections on the Motive Power of Heat* (Paris, 1824) is traditionally understood to be one of the foundation stones of thermodynamics, was an engineer trained in the Ecole Polytechnique; and his intention in publishing was not to contribute to the advancement of physics, but to promote the improvement and use of steam engines for the public benefit. 'The study of these engines,' he wrote, 'is of the greatest interest, their importance is enormous, their use is continually increasing, and they seem destined to produce a great revolution in the civilized world.'

There are many contemporary examples of this relationship between science and technology. The detection system set up by the United States government to monitor compliance with the Nuclear Test Ban Treaty of 1963 is one of these. Seismometers installed to detect underground nuclear explosions have produced a vast amount of new information about earthquakes; and satellites placed in orbit with instrumentation to detect bursts of radiation from nuclear tests have provided information about gamma radiation from astronomical objects, and led to the quite unexpected discovery of the occurrence of lightning strokes that are more than one hundred times more powerful than ordinary lightning.[21]

Perhaps the most dramatic example is provided by radio-astronomy. Here it could be argued that a whole new branch of pure science was made possible by technology, for it originated in a piece of research that was undertaken solely for technological reasons and to serve the interests of industrial enterprise: an investigation of the sources of the background noise that interfered with terrestrial wireless communication. This work was done at the Bell Laboratories in the 1930s. The listening instrument built by Karl Jansky, a physicist employed by Bell, was the forerunner of the modern radio-telescope, and the information about the background noise that he obtained by this means led to a remarkable discovery: that a major source of noise, a steady hiss, originated far outside the solar system, in the direction of the centre of our galaxy. The knowledge that has been generated as a result of this accidental yet momentous discovery is helping to transform the picture of the universe we inhabit, and hence our picture of ourselves. Some might interpret this story as a vindication of Engels's famous statement that 'if a society has a technical need, that helps science forward more than ten universities.' But to do so would be highly misleading. In fact, the story suggests precisely the opposite. It is true that a technical need existed, in that the Bell Telephone Company wished to improve the quality of radiotelephonic communication; but the possibility of radiotelephony itself was already entailed by the work of Clerk Maxwell, Hertz, and others who were grappling with purely scientific questions;[22] and the solution of the problems posed by this particular 'technical need' not only had scientific components but also presupposed an already existing body of background knowledge about the nature of electromagnetic radiation. Had this not existed there could have been no recognition that the 'technical need' might be satisfied by the application of science.

THE CONNECTIONS BETWEEN SCIENCE AND TECHNOLOGY IN SOCIAL THOUGHT

The implication of Engels's statement – that technology is the prime mover of science – is now widely believed. It is often linked to two other theses. The first is that the rise of modern science is causally related to the rise of industrial capitalism; or, as Engels also said, that the bourgeoisie had need of science and science arose together with the bourgeoisie. The second is that modern science – 'science since Galileo' – is inherently technological.[23] This thesis is advanced by, among others, the influential social theorist Jürgen Habermas and the ideologists of the Communist Party of the Soviet Union. Both draw their inspiration on this point from Marx: in the case of Habermas, via the thought of the Frankfurt School, especially Marcuse;[24] and in the case of the Soviet ideologists, via Lenin. This is not to say that only Marxists or the Marxist-inspired adhere to these propositions; on the contrary, they are shared by many intellectuals who have no necessary sympathy with Marxism, including many contemporary historians and sociologists of science.

The first thesis – that technology motivates science – has, I hope, been adequately disposed of in the preceding section. The third, the defence of which seems to require much abstract and dogmatic, not to say impenetrable, philosophizing, I shall refer to briefly in chapter 6. The second – that modern science and industrial capitalism are in some way causally related – I shall discuss now. This thesis has two interdependent components: first, that the science we know today is qualitatively different from the science of antiquity, and is a specific historical project beginning around the sixteenth century in Western Europe; and second, that this project could only have originated in the economic, social, and political conditions that were made possible by the capitalist mode of production.

1 / My contention is that, in the sense intended here, there is no such thing as 'modern' science. Contemporary 'externalist' accounts of the history of science seek, quite rightly, to set the science of a given epoch in its economic, social, and cultural milieu (which the older, 'internalist' accounts[25] conspicuously failed to do); and they warn us that we must avoid seeing and judging what was done in the past through late-twentieth-century eyes. At the same time, however, if we see the science of 'then' as something locally and culturally determined, as something transfixed within a context, understandable purely in its own terms (as many externalists do), and the history of science as nothing but a sequence of such incommensurable paradigms (Kuhn 1962), then our historical understanding will be as defective as the internalist, 'Whiggish' version[26] that it replaces; for we shall have ignored the fact that the science of 'then' began from, and was rooted in, problems that were raised by the science of 'before then.' 'It is not sufficent,' the historian Herbert Butterfield wrote, 'to read Galileo with the eyes of the 20th century or to interpret him in modern terms' (1957: 9). But, he added, 'we can only understand his work *if we know something of the system that he was attacking*' (emphasis added). Thus, for example, it makes no sense historically for Habermas to claim (1976: 156) that 'the mechanics of Galileo and his contemporaries dissects nature with reference to a form of technical domination which had just been developed within the framework of the new modes of manufacture' (that is, capitalism). It makes no sense because Galileo's 'problem situation' arose from, and was partly rooted in, earlier thought. This is not to say that the economic conditions of Galileo's time had no relevance, any more than it would be sensible to suppose that the state of the Roman church at that time had no relevance; it is simply to say that there is also an internal continuity in the unfolding

of scientific thought that cannot be ignored. As Butterfield says (p. 7): 'The so-called "scientific revolution" popularly associated with the 16th and 17th centuries [reaches] back in an unmistakably continuous line to a period much earlier still.'

In fact, if our conception of the pursuit of scientific understanding is disentangled from a fixation with its contingent instrumental possibilities – as it should be – one must see science as embodying a continuous tradition from, at the latest, the pre-Socratic Greeks. There is nothing that fundamentally distinguishes, for example, the Alexandrian Eratosthenes' calculation (circa 240 BC) of the circumference of the earth (to within about a thousand miles of its present-day approximation) from Galileo's correct deduction from his telescopic observation of sunspots,[27] in 1610, that the sun rotates once in roughly 27 days; nor either of these remarkable examples of inference and calculation from the work of any modern astronomer. If scientific thought is 'better' today, it is so only in the sense that the thinkers have more sensitive instruments, use more refined techniques, and have a much larger body of accumulated background knowledge on which to draw. The history of science, properly understood, tells us that scientific revolutions, like political revolutions, never constitute a complete break with the past and never foreshadow all the consequences that follow in the era that they usher in.

Copernicus's *De Revolutionibus Orbium Caelestium* and Vesalius's *De Humani Corporis* were published in the course of one wonderful week in the year 1543, and it has been said with little exaggeration that these two books brought the Middle Ages to an end. Yet both were rooted in an earlier tradition. Copernicus's revolving 'orbs' were not the planets themselves but the ancient celestial spheres in which the planets were supposed to be fixed; Vesalius's anatomical work falsified many of the generally accepted ideas of the Greek-born Galen who lived in the second century AD, yet he remained in many respects a Galenian. Kepler brought about a new understanding of the movement of the planets, and he advanced many other important ideas also (for example, relating to the physiology of vision); yet he believed his crowning achievement to have been the proof of the divine harmony of the spheres. Newton's secret obsession with alchemy led him to devote a large part of his life to the doing of alchemical experiments, some of which, for a time, seriously affected his health. This obsession, and his interest in the occult generally, led Maynard Keynes to characterize him as 'the last great mind which looked out on the visible and intellectual world with the same eyes as those who began to build our intellectual inheritance rather less than

10,000 years ago' (J.M. Keynes, 1972: 363–4). Charles Darwin was the Newton of the living world, yet even his fertile and inventive mind could not entirely shake itself free from the Lamarckian legacy of the inheritance of acquired characteristics.

2 / The notion that 'modern' science could only have originated in the conditions of emergent capitalism (the second component of the thesis I am examining) owes much to the Baconian legend. Francis Bacon, whose life coincided with that of Galileo, is widely claimed to have been 'the father of modern science.' This belief was carefully fostered by the founders of the Royal Society of London and was later accepted and propagated by the *philosophes* of the eighteenth-century Enlightenment and, in the early nineteenth century, by scientists like John Herschel. It is an extremely doubtful proposition; but its importance here lies in the fact that Bacon's 'setting of science on a new foundation' (as he claimed to have done) reflected ideas that could be said to coincide with the needs of the new economic order. His espousal of science for 'the relief of Man's Estate,' of practical experiment, of 'works of Fruit,' and his claim that science should aim to master nature, can be called in evidence on this point. The Baconian legend is that his program captured the spirit of a new, or reconstructed, science. The legend underlies the already mentioned remark of John Dewey that technologies 'gave that common-sense knowledge of nature out of which [modern] science takes its origin' as well as Habermas's piece of special pleading that 'it is only within the framework of modern bourgeois society ... that science can receive impulses from the experiential realm of manual crafts, and research can gradually be integrated into the labour process' (1976: 156).

According to Paulo Rossi, one of the most eminent of modern Bacon scholars, Bacon wanted to see a new science of collaborative, open, and free inquiry, experimentation, and unprejudiced observation applied to the practical arts of the craftsman. His goal was the marriage of a new science of natural history[28] (the directed and methodical classification and ordering of the 'facts' of nature) to the traditional 'erratic' experimentation and chance observation of the artisans, so as to form a single systematic corpus of knowledge (Rossi 1973). It was the new science, not the crafts, that was to benefit most from this. His contention, in the Preface to *The Great Instauration*, was that the mechanical arts were 'continually growing and becoming more perfect' while 'philosophy and the intellectual sciences' stood like statues 'worshipped and celebrated but not moved or advanced.' In his essay *In Praise of Learning* he wrote that 'all the philosophy of nature which is now received is either the philosophy of the Grecians or that other of the Alchemists,' remarking sourly that 'the one never faileth to multiply words, and the other ever faileth to multiply gold.'

In one sense Bacon *was* a liberator. He prefigured the birth of *organized* science[29] in, for example, the growth of scientific societies, the development of new canons of scientific discourse (the abandonment of secrecy, such as had been practised by the alchemists, in favour of an open exchange of information and ideas), and the secularization of science[30] (the removal of its theological-metaphysical elements). Yet Bacon's arguments for experiment were not motivated by a desire to test speculative theories, and his conception of observation was closer to that of the naturalist than to that of the true scientist. Bacon was more impressed by voyages of discovery, since these produced new and 'curious' facts, than he was by theory (he refused to accept Copernicus's theory, for example). He applied to his new science the same basic idea that he brought to a consideration of the law (it is sometimes forgotten that Bacon was also a lawyer). He was utterly hostile to general systems of thought (which he associated with the Schoolmen) and brought to his conception of scientific method the same inductivism that he wished to see applied to the law. Bacon thought that jurisprudence should consist of reasoning from particulars. The source of legal generalizations should be statutes and decided cases, not deductive reasoning from abstract principles. In Baconian language, legal generalizations should not be of the highest order (generalizations about the nature of justice) but of the middle order, for these were more fruitful (Shapiro 1969).

Thus, though Bacon's ideas were in one respect liberating, in another sense they would have pointed science in the wrong direction. He had a great influence on the 'projectors' of the Puritan Revolution of the 1640s and 1650s, among whom Samuel Hartlib was a leading figure, as well as on the members of the Royal Society following the restoration of the monarchy. All these men reflected an optimistic belief in material progress, a strong technological interest, and a passion for collecting useful 'facts'; but a quite different movement was growing elsewhere in Europe, especially in France, emphasizing problems and abstract ideas – especially those ideas that could be given shape mathematically; and this movement represents far better than Baconianism the continuing tradition of science that had been handed down from the Greeks. As Kuhn has suggested, only Newton perhaps among figures of the first rank in seventeenth-century England can be said to truly belong in that tradition. Keynes called Newton, in the famous passage referred to earlier (1972: 363), 'the last of the magicians.' He has also been dubbed 'the first of the *illuministi.*' He was neither. Rather he was, as G.L. Huxley once said, 'the greatest of the Hellenic geometers.'

A reading of the work of modern Bacon scholars cannot fail to convince us of the falsity of the picture of him as 'the father of modern science.' The idea of science as inherently technological is what he preached – that and the gospel of a future technological society. He was, as the Marxist Benjamin Farrington (intentionally anachronistically) argued, 'the philosopher of industrial science' (1973). In fact, Baconian ideas had little effect on the established sciences like astronomy, optics, and mechanics (Kuhn 1977: 136–7), though they did perhaps help to open the way for the later investigation of new phenomena like magnetism, electricity, and heat. Nevertheless, in spite of the evident lack of impact of Baconian ideas on the actual progress of science, they came to support an ideology that identifies 'modern' science with the existence of capitalistic modes of economic organization.[31]

THE 'LOGIC' OF TECHNOLOGY AND THE 'LOGIC' OF SCIENCE

The crucial fact about science is that its product, scientific knowledge, exists outside and independently of the world of practice, though it may, and frequently does, enter into the world of practice. Scientists produce propositions about the natural world; and although these propositions may have implications for practical action (whether acted upon or not), they are not themselves part of the world of practical action. Once created they have an autonomous existence. Parts of scientific knowledge may be found to have practical application and parts of it may not; but practical application is neither a necessary nor a sufficient condition for its creation – or, for that matter (contrary to what some theorists believe) for its truth. The latter can be seen in the fact that a scientific theory may be successful in practical application and yet be false. Newton's laws are now known to be false (though they contain some truth),[32] but they are highly successful in practical application to space flight. It is true that the products of technology – tools, instruments, and systems designed for their use – embody ideas, and these certainly can be said to exist independently of the products themselves, in the sense, for example, that we can look at and examine critically the ideas embodied in the technologies of other cultures, past and present, and indeed, at those embodied in our own technology; but the products and the use of technology per se exist entirely within the realm of practice. Their sole purpose is to alter and control the material conditions of life. The products of science, which are ideas, theories, problems, exist as such, wholly outside the realm of practice. Where scientific knowledge is actually applied it becomes technological knowledge. The fact that science and technology nowadays interact in complex ways, even as Barnes has argued (1982b: 166) 'interpenetrate totally,' does not eliminate the differences between them. Nor does the often-mentioned fact that the two activities of doing science and pursuing technology are frequently combined in the same person. Archimedes, Galileo, and (in our own time) Bethe, Fermi, Feynman, and others who worked on the atomic bomb, all contributed to technology; but that does not make them technologists.

There are connecting principles in the progress of science that are absent from the progress of technology (Polanyi 1951: 74–5). Polanyi's example was the technological progression: the rushlight, the candle, the kerosene lamp, the gaslight, the electric arc lamp, the enclosed incandescent lamp, the mercury and sodium lamps. All these constituted improvements in illumination (as well, incidentally, as a movement from simple craft-based technology to science-based technology), but each was simply a moment in a series of logically unconnected solutions to a succession of practical problems. In science, however, we do find connecting principles; for example, in the progression from Kepler, through Galileo and Newton, to Einstein (one could even begin with Anaximander and Aristarchus). Here one can trace a connected movement of ideas, an unfolding of thought.

The fundamental aim of science is to propose theories that assist our understanding of natural events and processes, and to test these theories as rigorously as possible. The latter is crucial. The search for empirical (test) evidence is instrumental for the scientist. For the technologist, however, it is the theory that is instrumental; and this is true whether the technological theory is based on science or, as in traditional technology, on rule of thumb, canons of accepted procedure, or even myth.[33] What is important for the technologist is the specification of the initial conditions within which the theory or law is taken to apply, and the predicted outcome. If an engineer, for example, wishes to build a bridge that will stand up to specified stresses he will look for the conditions which, taking certain 'laws of nature' for granted, will produce the desired result (leaving a large safety margin for error in the specification of the conditions). He is not interested in explanations; nor is he interested in discovering new and surprising effects, as is the scientist, and he will, in some circumstances, even discount them if he comes across them. Nor is he concerned to put forward and test more inclusive theories. His focus of interest is different because his aims are different. The modern technologist needs to know a lot of science, though it is chiefly the science that is taught from the textbooks (see above); but he is not concerned about the advancement of science except in so far as this is necessary for the pursuit of his specific problem-oriented needs. When the technologist uses theories, he uses them as tools.

Scientists seek true knowledge, even though it has long been accepted (by most of them) that their propositions can never be known to be certainly true; they accept that today's science is subject to revision and improvement. Technologists seek only knowledge that is likely to lead to a sufficiently successful outcome to serve their purpose; and they may sometimes make use of knowledge that is known to be false. Mario Bunge (1966: 329) gives as an example the design of optical instruments, which makes use very largely of what was known about light in the seventeenth century.

Technology is concerned with design. The ends of a piece of technology must be deliberately and carefully specified. A technology must 'declare itself in favour of a definite set of advantages and tell people what to do in order to secure them' (Polanyi 1958: 176). Technology, unlike science, is not concerned with things as they are but with things as they might be. 'In science we investigate ... reality; in technology we create a reality according to our design' (Skolimowski 1966: 374). Technology – or at any rate advanced technology – is in no way inferior to science in ingenuity, creativity, or intellectual challenge; but in science these qualities are directed towards a deeper understanding of the nature of things, whereas in technology they are directed towards turning known facts to positive advantage.[34]

Science must progress in order to exist, and it progresses through the increase of knowledge and, particularly, by increasing the *depth* of knowledge about the natural world.[35] Progress in technology, by contrast is measured in terms of effectiveness – effectiveness in producing objects of a specified kind, new and 'better' objects, 'better' being measured by improvements in durability and reliability, and lower costs (Skolimowski 1966: 375). The criteria of technological progress cannot be replaced by, or even be meaningfully translated into, the criteria of scientific progress (ibid., p. 374). A scientific proposition may lead to an advance in understanding – scientific progress – even if it has no technological application, and a piece of technology may have great practical value and mark a real step forward even if it adds nothing at all to science

(Polanyi 1951: 73–4). Moreover, it is of no consequence to pure science whether a discovery has a use – though its utility may be an uncovenanted benefit and bring some prestige. In advancing technology we may advance science; and in advancing science we may advance technology; but this is a contingent, not a necessary, relationship.

3

Science and the State

I have done the state some service, and they know it.
Othello, v, ii, 338

It is a truism that the practical activity of doing science has undergone great changes in the past fifty years. Even as late as the end of the third decade of this century, the scale of scientific research was still relatively small, scientists constituted a rather tiny segment of the professional work-forces of the industrialized world, their business was conducted for the most part in obscurity, and the funds available for it were minuscule by present standards. Today we live in an era of Big Science: science done by huge numbers of scientists, often in large laboratories employing teams of workers, using immense amounts of sophisticated equipment, producing a vast annual output of scientific papers (for example, some 350,000 published each year in about 12,000 journals in chemistry alone).[1] It is true that the majority (international statistics usually show about 85 per cent) of all those officially classified as scientists work on technological applications of science. Moreover, much of the work done by the other 15 per cent is of a relatively routine character. The real breakthroughs in all the sciences are made by probably no more than 5 or 6 per cent of all scientists, with the remaining 9 or 10 per cent in this category engaged in follow-up research. There are sciences, and indeed countries, in which small groups, and even individuals working with one or two technicians, are still the norm. The same applies to the scale and sophistication of the equipment used. Theoretical physicists still work with chalk and blackboards, though increasingly with computer terminals also (in contrast to their experimental colleagues who must bid for scarce time with the latest particle accelerator or large optical or radio telescope). Palaeoanthropologists grub about in the soil with simple tools and brushes as they have always done.

Nevertheless, while it is misleading to think of all science today as Big Science, the fact remains that it is now a complex collective enterprise and pervasive in its influence. It is no longer done by an underprivileged group working away from the public eye, relegated to the basements of university buildings where it was thought they would be least likely to interfere with the pursuit of genuine scholarship. Scientists have taken their place as an influential force in society. At the same time, the state has emerged as the chief sponsor and promoter of scientific research. It has become a major purchaser of scientific knowledge and techniques for military purposes and for the pursuit of 'national welfare.' In a world where governments demand their services more and more, and where they can claim a large share of public funds to support their work – in a word, where science has become a 'national asset' – it has become virtually impossible for scientists to remain unpoliticized or to resist insistent demands for 'social accountability' and claims on their allegiance to the pursuit of national security and national welfare.

Scientists themselves must bear much of the responsibility for fostering the image of the practical usefulness of science.[2] Today, the association of the production of new scientific knowledge with technological application has become very close. We noted, in chapter 2, examples of scientific discoveries that led eventually to technological advance even though this was far from the minds of the scientists concerned when they embarked on the research: an *unintended consequence* of pure science. We also noted that much pure science is now done in general support of 'missions' – research that is needed for the successful outcome of a mission prescribed by government and not directly by scientists themselves. Such research, though motivated by an *intended application*, possesses an intrinsic interest in enlarging understanding that would exist even if the research were not oriented towards the accomplishment of a practical purpose. (An example is some at least of the theoretical physics that was done at Los Alamos between 1943 and 1945.) One might call this 'pre-technological science.' It is pure science with clearly discernible potential application to technology, and it is supported by the state for this reason. Increasingly, however, governments underwrite pure science that is not undertaken for an intended application, not because it is valued 'for its own sake,' but because it may conceivably have some future application.

The importance of science to society and to the state has not been achieved without cost to its practitioners. Among the costs that scientists have been called upon to bear are those that relate to two of the traditionally most cherished aspects of the scientific ethos: the freedom to pursue knowledge unhampered by interference from authorities within the society in which they work, and the freedom to communicate their ideas without hindrance to the international community of scientists to which they belong.

THE BUREAUCRATIZATION OF SCIENCE

But it is pretty to see what money will do.
Samuel Pepys, *Diary*, 21 March 1667

Today, little science and big science are alike subject to a measure of political and bureaucratic control that is unparalleled in the history of science. In authoritarian states like the Soviet Union all scientific work is potentially subject to the scrutiny and direction of state and party bureaucrats.[3] This could be said to be symbolized in the fact that the All-Union Academy of Sciences, the highest professional body of scientists in the Soviet Union, has the status and responsibilities of a government department. In this it is quite unlike the academies of science in any western country. Soviet scientists retain a precarious degree of autonomy that depends for the most part upon the influence that their leaders are willing, or are allowed, to exercise in the higher councils of the state. Some autonomy there must be under any political regime, however, for no matter how closely the scientists' tasks may be regulated by external criteria, they must also obey the constraints that nature imposes. As Bacon said, nature to be commanded must be obeyed. Most governments, including the present government of the Soviet Union, recognize this imperative. When they do not, as in the case of genetics in the Soviet Union under Stalin,[4] the result can be disastrous for the state as well as for science. Nevertheless, the extent of control and direction of science is much greater in the Soviet Union than it is in the West, and Soviet scientists are continually exposed to the possibility of ideological supervision by party cadres. Moreover, Soviet science is planned to a degree that would not be tolerated in the West.[5] In the West, the bureaucratization of science has come about in large measure as a result of its cost in national resources and as a consequence (some of it unintended) of the operation of the funding systems employed. Scientific progress, and the ambitions of scientists, have made the enterprise more and more costly; where governments have been willing to meet these costs it is because they have found in science an indispensable asset.

There are a variety of mechanisms in use for allocating financial support, and different countries use different mixtures. Grants may be made to support the research of individual scientists or the leaders of research teams; they may support specific projects or a general area of research; they may be made to a university or to a university department to be used as the university or department decides; they may be given to build on existing strengths in some disciplines or to make good deficiencies in strength in others; they may be made for one year or for a number of years. The granting agencies are usually independent bodies to which the government allocates a block of money that is then distributed at the granting agency's discretion. The purpose of this device historically, and in principle, is to insulate support for science from direct political interference. Examples of such 'buffer' agencies are the National Institutes of Health (in the United States), the Natural Science and Engineering Research Council (in Canada), and the Environmental Research Council (in Britain). In addition, there is almost always direct state financing through research contracts let by government departments and similar public agencies. Contracts are normally (though not always) more specific than grants as to the nature of the work to be done, the date by which the research must be completed, and so on.

An increasing number of American universities now employ professional lobbyists whose job it is to persuade Congress to vote funds directly, especially for buildings and equipment (that is to say, they seek to circumvent the regular funding system). This strategy reflects the straitened circumstances in which many smaller and less prestigious universities now find themselves. The amount of money involved increased from $3 million in 1982 to $137 million in 1985. The practice is criticized by many scientists and science administrators because it invites log-rolling and removes decision-making from the expert scrutiny of scientific peers. This scrutiny is intended to temper scientific autonomy with a measure of public accountability, and is formalized through advisory committees and panels of working scientists drawn from the appropriate disciplines and subdisciplines. They evaluate research proposals and are expected to advise the granting agencies on such questions as the competence of the scientists who will conduct the research, and whether the proposed project is of sufficient scientific interest to warrant expenditure of the amount of money requested. Such questions are aimed at judging how likely it is that a given allocation of resources will yield positive scientific results. They can only be answered – even if imperfectly – by scientists. In the case of pure science done under contract, which is usually pre-technological science (as previously defined), external criteria such as urgency and degree of social relevance may be applied as well.

Whatever the precise arrangements, those charged with making allocatory decisions, and the scientists who advise them, inevitably exert a substantial de facto influence on the quality and differential rates of growth of research in specific disciplines, and thus on the quality and growth of the nation's scientific research as a whole. They bear a heavy burden of responsibility, since mistakes in directing science will have lasting effects. For example, one of the criteria that has commonly been applied in assessing grant proposals is whether the project is 'of exceptional timeliness and promise.' The trouble with this standard is that many original ideas appear at first to be both unpromising and untimely – which is another way of saying that if Darwin were alive today he would probably fail to get a grant for his work on bean shoots. Florence Nightingale, it is said, was determined that, if hospitals cured no one, they should at least not be permitted to spread disease. An analogous principle could be said to apply to those who provide public support for science: if they cannot advance science they should at least try not to impede it. Unfortunately, this is more easily said than done.

The past practice in many countries (which has been particularly important for pure science) of providing a substantial portion of state aid in the form of unconditional block grants to universities, laboratories, or university departments avoids some of the defects that are apparent in other methods of funding (Roy 1984).[6] It also gives these scientific institutions greater control over their own destiny. Unfortunately, those governments (like the British) that have successfully used this method in the past are now undermining it by seeking to impose conditions on the way the block grants are to be used.[7] There is also a growing tendency for governments to intervene more directly in the allocatory process by bringing pressure to bear on 'buffer' agencies to pay closer attention to government preferences. For example, in 1982 Dr George Keyworth, then President Reagan's chief scientific adviser, and a political appointee, stated publicly that it was 'eminently reasonable' for the president to want granting agencies like the National Science Foundation to 'share his political philosophy.'[8] The incident caused a scandal; but it was clearly a straw in the wind.

Since the funding systems in use differ in detail from country to country it is not easy to generalize; but some of the consequences of the operation of these systems should be noted. Some granting agencies require the submission of extremely detailed research proposals and may even demand that the expected results of the research be specified. An immense amount of the time and energy of scientists is thus diverted from their proper function. If the application fails, the time spent on preparing it is wasted. Such grant applications are, moreover, 'fraudulent' in a non-venal sense. As Ziman has written (1987: 97), scientists engaged in pure research can usually only state their intentions – the outcome cannot be known in advance. Yet public accountability demands that granting agencies 'are bound to press grant proposals into a mould of superficially formal rationality.' The system also tends to encourage actual dishonesty. Applicants may disguise their real intentions, seeking to increase the political attractiveness of what they want to do. 'The incentive to oversell [pure] research on grounds of its utility becomes greater as the competition for research funds becomes keener' (Brooks 1978: 181). It is common for scientists to apply for grants for work that is nearly completed. If successful the grant will then be used towards the next project. This device of 'rolling over' funds is rationalized by the argument that it reduces uncertainty.[9] The need to negotiate grant applications has given rise to the practice of 'grantsmanship' – the development of sophisticated techniques for persuading granting agencies – and, in the United States at least, there are now commercial organizations that specialize in assisting applicants for a fee. As Dr Harvey Brooks has said, 'The capacity of the scientific community to subvert attempted political management is much greater than in other fields of endeavour [because of the "esoteric" nature of science] but the more such subversion is necessitated by ill-advised political management, the greater the mistrust of scientists' (1978: 183). It is thus of prime importance, he says, that the mechanisms governing the relationship between science and the state should not place too great a strain on the intellectual honesty of scientists. 'If the cost of being honest about ... one's research is the total loss of support, then a kind of Gresham's Law will set in which rewards those scientists most willing to stretch the bounds of intellectual honesty in explaining the benefits of their research.'

The agencies providing support for science also require post facto reporting, and the pressure on the scientist (especially the young scientist) to prove himself or herself to the agency may lead to hasty and slipshod work; to the distortion of results sometimes, especially in contract research, to bring them into line with what the researcher believes the agency would wish them to be; and even (on occasion) to outright fraud. There have been a number of well-publicized instances of the latter in recent years (one of them, the Summerlin case, is referred to in chapter 5) leading to disgrace and the ruining of careers; and many scientists have testified that there is a good deal of minor fraud that is never publicly revealed. John Ziman has suggested that the competition for grants, and the pressure for results that will make possible further grants, tend to foster a conservative attitude towards risk-taking in the average scientist's research plans. He says that 'the majority of scientific careers give the impression that minimal risk is their underlying policy' (Ziman 1981: 15). Competition and pressure for results may also be a factor in attracting young scientists to a career in Big Science, where they can work in laboratory teams and the risk is shared. This is not, of course, to suggest that all 'little' scientists are adventurous risk-takers and all 'big' scientists are conservative risk-minimizers. There are, as Ziman concedes, 'Dionysian' scientists in Big Science and 'Apollonian' scientists in Little Science.[10]

'Dionysian' scientists are the leaders in advancing science. They have wide interests beyond their own specialism. They are 'continually on the alert for the intimations of discovery ... Their research policy is not to tie themselves down for life to a single major problem, not to get caught in the web of routine solutions of puzzles, but to keep themselves as free as possible to exploit any ... opportunity presented by experience or reflection' (Ziman 1981: 17). Many of them have 'the kind of instinct which guides a scientist to the right problem at the right time before the rightness of either can be demonstrated' (Ziman 1981: 17, quoting Sayre 1975). By contrast, Apollonian scientists are content, as he puts it, to 'make a modest living.' They may, indeed, be inclined to design their research proposals 'as closely as possible according to some fashionable model.' Such conduct, Ziman says, is strongly encouraged by recent economic pressures on universities. It has become common, for example, for universities advertising for applicants to fill scientific positions to specify that the successful candidate will have 'proven ability to generate grant support.'

Many scientists now agree, privately and sometimes publicly, that much of the science that is now done at great expense is justified neither by its quality nor by the significance of the results. One recent commentator, Nicholas Rescher, claimed to have established statistically that there is actually a diminishing return on scientific effort; that first-rate science does not even increase proportionately to the overall growth of scientific output. Scientific progress depends upon originality, but as Arthur Kornberg, a Nobel Prize–winning biochemist, has said,[11] 'we don't have as large a fraction within the scientific community [today] of people whose prime, driving and overriding interest is the pursuit of basic knowledge. [Today's scientists] are not primarily interested in science as an art form.' The ruthless competitiveness that exists in many branches of science today – the 'rush to the chace with hooves unclean, and Babel clamour of the sty'[12] – is not simply competition for public funds. It is also competition for recognition by peers who are now far more numerous than they once were; it is competition to get ahead of rivals in making discoveries; it is a struggle, even, to obtain a permanent foothold in the profession. One older scientist put it this way: 'Everywhere in science the talk is of winners, patents, pressures, money, no money, the rat race ... things that are so completely alien to my belief ... that I no longer know whether I can be classified as a modern scientist or as an example of a beast on the way to extinction' (Goodfield 1981: 213). The ills that now beset science cannot be attributed solely to bureaucratic interference with the orientation and conduct of research that is a consequence of state sponsorship. But with money must come accountability. Lord Rutherford, one of the founders of modern physics, is reputed to have told his younger colleagues at the Cavendish Laboratory sometime in the 1920s: 'We haven't got the money, so we've got to think.' Now, scientists have money in amounts that would have been inconceivable to their predecessors half a century ago. It has taken them little more than a generation to discover the truth in Oscar Wilde's observation that, while it is disagreeable to be frustrated in life, the real disasters begin when you get what you want.

THE DECLINE OF THE INTERNATIONAL COMMUNITY

The hardest bones, containing the richest marrow, can be conquered only by a unified crunching of all teeth of all dogs.
Franz Kafka

The classical picture of scientific practice, as portrayed by Michael Polanyi (1962), was of a kind of free-market economy. Each scientist acted independently, yet because all scientists adjusted their activities to what others were doing, the interests of the whole were achieved with maximum efficiency. This 'hidden hand' not only operated within a national scientific community, it worked also across national boundaries. There was a world scientific community consisting of all those persons who, irrespective of nationality, race, or creed, shared the scientific ethos, spoke the same scientific 'language,' followed the same scientific traditions, and communicated freely, receiving and assimilating the results of research wherever it was done. The implication was that there could be no such thing as American physics, German chemistry, Russian geology, or Canadian biology; there could only be physics, chemistry, geology, and biology. It took a Hitler to decree German physics, and a Stalin to decree Russian biology – in both instances, with disastrous results. This universalism, or supranationalism, could be explained not only by the fact that science could not be done in isolation, or only because it was in its nature an open, collaborative enterprise, but also by the fact that the more widespread the community the swifter the progress of science could be. The microbiologist Salvador Luria has written: 'Science is an immensely supportive activity ... The reassurance that a physicist gets from knowing that every colleague the world over believes in the correctness of Maxwell's or Boltzmann's equations, or a biologist from knowing that all biologists [accept] the structure of the DNA molecule, is not just intellectually reassuring, it is also emotionally satisfying because it implies a sharing of knowledge and membership in a segment of humanity that speaks and thinks in a common language' (1984: 117). A scientist, says Zhores Medvedev, the Russian biologist who is now in exile, 'sets out on a course of international cooperation, not under the influence of some directive, but on account of a natural need, a natural necessity connected with his current scientific work, with the more rapid solution of some problem or question in mind. The very structure and burgeoning activity of the world community of scientists involve the individual scientist in scientific cooperation, drawing him like a grain of sand into a whirlwind' (1971: 6).[13]

This universalistic goal of science, which was largely achieved in earlier times, and is still held in respect by scientists, is regrettably more and more limited today by the particularity of the world as it is. It is true that the history of science provides many instances of national competitiveness, rivalries, and jealousies, for example between French and British scientists in the seventeenth and eighteenth centuries over the issue of Cartesian versus Newtonian physics. It is also true that the ostracizing of German scientists by their British and French colleagues at the end of the First World War (the Germans were, for instance, banned for some years from international conferences) was due not simply to the intense emotions that the war had aroused,[14] but also to envy of the preeminent place that German science had attained during the nineteenth century. Germany had become the Mecca to which most aspiring scientists went for their postgraduate work, and by 1914 roughly half the world's scientific journals were published in German in the German language. Discreditably, many allied scientists saw in the humiliation of their German colleagues an opportunity to redress the balance.[15] The absurdity of this affair in the history of science is symbolized in two post-war incidents. First, Fritz Haber, the scientist who had directed German poison gas research, was named in the Versailles Treaty as a war criminal, yet in the same year (1919) he was awarded a Nobel Prize for his pre-war scientific work in chemistry. Second, shortly after the Armistice, Einstein had to refuse an invitation to address the French Academy of Sciences because he had been advised that members had threatened to walk out if he appeared. This in spite of the fact that Einstein was totally opposed to the war, and had been strongly critical of the part that Germany played in it.

Nevertheless, history also records examples of scientific needs overcoming discord between nations. Frederick the Great of Prussia appointed a Frenchman, Pierre-Louis Maupertuis, as president of the Prussian Academy of Science and had the transactions of the academy written in French on the ground that French was more universally understood than German – a practice that was continued during the Seven Years' War even after Maupertuis's death in 1759. In 1778 the fighting between two contingents of British and American troops in the Revolutionary War was halted for three days to allow an expedition of 'rebel' American scientists to pass through the British lines with telescopes to observe an eclipse of the sun; and a year later Benjamin Franklin, then the American secretary of state and himself a scientist of some note, issued a remarkable instruction to the captains of all armed ships at war with Britain not to interfere with the passage of Captain Cook's vessel, which was then on a voyage of discovery to 'unknown seas.' His message concluded: 'This, then, is to recommend to you that should the said ship fall into your hands, you would not consider her as an enemy, not suffer any plunder of the effects contained in her, nor obstruct her immediate return to England.' During the Anglo-French ('Napoleonic') wars, the English chemist Sir Humphry Davy was given a safe-conduct pass so that he might visit Paris to address a meeting of the Institut Français. Such incidents are likely to be rare today.

A striking example from another war, which perhaps symbolizes the threat under which the communality of science must nowadays suffer, is an incident that occurred in 1943. In January of that year, two scientists hurried to Chicago from a secret laboratory in Montreal where they were working on the atomic bomb. The purpose of their journey was to receive from American scientists working in another secret laboratory on the campus of the University of Chicago, and on behalf of their Canadian, British, and French colleagues in Montreal, basic data about the world's first working nuclear reactor and four milligrams of plutonium. The occasion for this hurried visit was the impending suspension of all further atomic co-operation between American scientists and those of her allies on orders from the United States government. A few days later the same exchange of friendly services, so characteristic of science, would have constituted, for the Americans, an act of treason.[16]

Ideally, a truly international community of science would require that there be no restrictions on the publication of the results of research, wherever it was done, and none on the international movement of scientists. There should be no blocks to the free flow of information, and scientists should be able to move about the world without hindrance to meet with foreign colleagues, attend conferences, take short- or long-term appointments in other countries, and even emigrate. These requirements are of special concern to countries with small scientific communities, and in areas of science where the cost of providing sophisticated equipment is beyond the means of the country acting alone (high-energy physics is a good example, one that led to the creation of CERN, the European Centre for Nuclear Research, which is staffed by scientists from seventeen countries). Both requirements are subject to constraints of greater or lesser severity in all science countries today in the interests, real and sometimes alleged, of national defence and security and the conduct of foreign policy. There is nothing new about this in principle. It has long been true that some science – pre-technological pure science as well as applied science – is 'born secret'; work on chemical and biological weapons, for example. Such work is conducted under conditions of strict security in peace as well as war. Research results are seldom published, and collaboration with foreign colleagues is permitted, if at all, only under the most carefully controlled circumstances, and only, of course, between scientists of countries deemed friendly. An example is the Technical Cooperation Program in chemical and biological research that was begun during the Second World War under the terms of a secret agreement between the governments of the United States, Canada, Australia, and Britain. The program permitted the sharing of research results and the establishment of integrated research projects. But in the atmosphere of international tension that has existed for the past forty years, the principle has been extended well beyond research that is 'born secret.' Francis Bacon's scientists in the imaginary research institute that he called Salomon's House took 'all an Oath of Secrecy for the Concealing of those [Inventions and Experiences] which wee thinke to keepe Secrett' – that is to say, secret from the state.[17] Today the shoe is on the other foot. Governments increasingly seek to tell scientists which of their 'inventions and experiences' shall be made public; and they do so not merely for reasons of national security in the older and stricter sense of the term, but also for reasons of commercial advantage and economic dominance over other nations, and especially those nations that are considered potentially hostile.

It is important to recognize, however, that there are strict limits to the effectiveness of secrecy in science; and, again, I refer here specifically to pure science that is, or may become, the basis for technological advance. The pure science that was done in Los Alamos in support of the Manhattan District project was accomplished, like every other aspect of that project, under conditions of tight security; but this did not prevent the Soviet Union from developing its own atomic weapons program and successfully testing an atomic device within five years of the United States. It was frequently argued in the 1950s (for example, in the United States Congress) that this was made possible by failures in the security system, and especially by the information about the Manhattan District project that was given to the Russians during the war by Klaus Fuchs, the 'atom spy' who occupied a central position as a theoretical physicist at Los Alamos.[18] But this argument was never very convincing, and it is now public knowledge that it had no basis in fact. In 1938, in Germany, Hahn and Strassmann succeeded in fissioning uranium by bombarding it with neutrons. This result was published in the usual way, and it led in the early months of 1939 to a flurry of fission-related research in Europe and North America and the publication of more than one hundred papers. Since by that year it was clear that Hitler was bent on precipitating a general European war that he believed could be easily won, scientists in Britain, France, Germany, and the Soviet Union all warned their governments that these recent discoveries made an atomic weapon in principle possible. The relatively slow progress that the Soviet atomic weapons program made before 1945 was due in part to the bureaucratic and highly centralized conditions under which Soviet science was done; in part to the 'liquidation' of many scientists during the Stalinist terror of the 1930s; in part no doubt to Stalin's lack of interest in the project until 1945, when he acted decisively; but chiefly to the immensely destructive invasion of the Soviet Union by the German armies in 1941, which made an all-out effort impossible.[19] The chief value that the Soviet Union got from Fuchs was the information he provided about the progress that was being made at Los Alamos; that is, how close the scientists and engineers were getting to producing a workable weapon. It was Fuchs who told Soviet intelligence of the success of the Trinity Test in July 1945.[20]

Given a level of theoretical and experimental maturity in a science such that a point of possible application in technology has been reached, the attempt to frustrate potential enemies by keeping knowledge secret is likely to be effective only in the short run. It was commonly believed in America in the years immediately following the Second World War that atomic secrecy might be maintained indefinitely and that this would allow the United States to keep its monopoly of atomic weapons. No scientist believed this, but many publicists and politicians did. They also believed that the chief threat to the maintenance of secrecy was the scientists themselves (Shils 1984: 421). These beliefs were totally unfounded. The fact that a certain discovery has been made by researchers in one country may be kept from researchers in another country; but, given the nature of the scientific enterprise, there is no way by which the latter can be prevented from making the same discovery independently. The prevalence today of simultaneous discoveries, particularly in the 'frontier' regions of science, by two or more groups of scientists acting in ignorance of the others' progress testifies to this fact. Pure research in genetics is now classified in the Soviet Union. Soviet geneticists are not allowed to discuss their results with foreign colleagues unless the results have been officially approved for publication; nor, it can be assumed, can they circulate preprints of papers within the Soviet Union – a practice that is now common in science because of long delays in publication in the scientific journals. The reason for classification, presumably, is that much pure research in genetics is now pre-technological and, as such, has military potential (in the production of genetically engineered organisms that could be used in biological warfare). Given, however, that scientists in the West are as aware of this potential as are their Soviet counterparts, and possess the scientific capability to exploit it, one may ask what purpose this secrecy serves. The answer can only be that it is believed to give a temporary advantage.[21] As Joseph Cade, a scientist who worked in atomic research, has said: 'secrecy in science cannot in fact *deny* knowledge, but can only *delay* the generation of that same new knowledge somewhere else by withholding information, so that those from whom we would have secrets must go through the same creative metamorphosis to discover them' (1971). It ought to be plain, therefore, as a matter of state policy, that any advantage to be gained from the suppression of scientific knowledge should be balanced against its long-term effects on scientific progress. It is futile to believe that new fundamental knowledge can long be kept secret from any country that possesses a critical mass of competent scientists in the relevant discipline.

It is often argued (for example, by Bok 1984: 157–63) that scientists are being sophistical in opposing secrecy since they frequently practice it in their own work – for example, by withholding data from rivals and being deceptive about their progress with a project in order to throw their pursuers off the scent, or to conceal the fact that their research is running into difficulties. All this is true but beside the point. Though scientists may want to keep their work secret while they are doing it, sometimes for the thoroughly laudable reason that they want to make certain of their results, none would wish to keep it secret indefinitely, for then it would have no point. In pure research at least, there is a fundamental incompatibility between the need for scientific progress and the banning of the publication of scientific findings. A tragicomical episode reported in *Physical Review Letters* a few years ago is illuminating. The editors of that journal received a communication on the subject of laser-induced fusion, one of many different techniques under investigation as a possible solution to the problem of achieving controlled thermonuclear reactions for electric power production. The editors sent this communication to a referee in the usual way, and the referee prepared a long and critical report. He was, however, able to release only a small portion of it to the editors because the rest had first to be submitted for declassification. When the author was informed of this he replied that he was not surprised, and that the criticisms that the referee had made probably related to material that he had felt himself obliged to suppress because it was secret.

Attempts by the government of the United States during the 1980s to impose more stringent restrictions on the communication of scientific knowledge have been sharply criticized by scientists and university administrators. This is not a purely domestic matter. The pre-eminence of American science in most major fields of scientific endeavour today means that this issue is of vital importance for scientists everywhere. The restrictions were aimed primarily at controlling the flow of knowledge that may have military potential, and are largely directed against the Soviet

Union, but they were in part a response to growing fears that the United States has been losing its competitive edge internationally in high technology, especially to Japan.

There is no doubt that there has been a substantial clandestine export of high technology from the West, and especially from the United States, in recent years.[22] Scientists have not contested the seriousness of this problem. A panel appointed in 1982 by the National Academy of Sciences, which included former government officials and executives from high-technology companies as well as scientists, concluded, however, that 'very little' of this technology transfer could be attributed to open scientific communication per se.[23] Whether or not this conclusion is valid, scientists have taken their stand against growing restrictions, in part on the United States Constitution, which has been held judicially to forbid restraints on publication except in extreme circumstances;[24] but more pragmatically, because of grave doubts about the scope and probable interpretation of the government regulations that have been promulgated to enforce these restrictions. For example, in 1983 the United States Department of Commerce issued a draft revision of its Export Administration Regulations that contained a new category of exports labelled 'critical technical data' – that is, knowledge that was, though 'critical,' not classified secret. The export of such knowledge would require a licence, which might of course not be granted.

The National Science Foundation argued that these regulations would extend not only to research done in the United States in which foreign nationals participate, but also to the presentation of scientific papers by Americans at conferences in the United States attended by foreign nationals, to the presentation of scientific papers by American scientists at conferences held outside the United States, and to the submission of scientific papers to foreign journals – in other words, to a wide range of normal scientific activities.[25]

In October 1983, the U.S. House of Representatives adopted an amendment to a bill extending the life of the Export Administration Act 1979 (under which these draft regulations were made), which stated that 'it is the policy of the United States to sustain vigorous scientific enterprise. To do so requires protecting the ability of scientists ... to freely communicate their research findings by means of publication, teaching, conferences and other forms of scholarly exchange.' This text was altered in the Senate by inserting 'non-sensitive' before the words 'research findings,' which, of course, would have entrenched the principle of open-ended restriction to which scientists were objecting.[26] The invention by

U.S. federal agencies of new categories of non-secret research labelled 'sensitive' (Department of Defense), 'controlled' (Department of Energy), and the like (seemingly ushering in a new era of the non-secret secret), together with an executive order of the president in 1982 tightening the classification system itself and even authorizing the reclassification of material previously declassified,[27] is precisely what has alarmed scientists and has led them into confrontation with government officials.

As one commentator has said, secrecy is now negotiable. For example, in 1984 the presidents of three major American universities[28] jointly protested the Department of Defense regulation establishing the category of 'sensitive' research, and said that their institutions would refuse to accept research contracts that imposed pre-publication review on that ground. These were powerful and wealthy institutions, indispensable sources of research that the state urgently needs to be done. Others less favourably placed are less likely to resist encroaching governmental control. Moreover, new regulation has been buttressed by more stringent action within existing regulations. Staff members of the National Academy of Sciences have complained that officials have stretched their authority 'beyond its previous limits' (Shattuck 1984: 424). The result has been to create what the *Bulletin of the Atomic Scientists* called an atmosphere of 'intimidation leading to self-censorship in science.'[29] Certainly, these developments have created a mood of caution on the part of American university administrators and many have sought to anticipate government intervention by imposing their own restrictions first. Conferences have been cancelled in anticipation of official objection to the presence of foreign nationals, and some have been announced as open to United States citizens only. It is true that the most flagrant examples have so far been conferences on technological subjects, but there is nothing in principle to prevent the practice from being extended, in spite of frequent public assurances to the contrary, especially since, in many instances, technological content cannot be easily separated from the basic science that underpins it.

There has also been a substantial increase in the number of granting and contracting agencies that impose pre-publication review. This means that research results cannot be published until they have been submitted to a contract or grant officer of the agency concerned for clearance. The official may require the rewriting or deletion of material, or may specify that the paper not be published before a prescribed date. If the researcher refuses to comply with any of these conditions clearance may be withheld indefinitely. It has been objected that the requirement of pre-publication review is frequently included in contracts and grants as a matter of routine rather than for a substantial reason. Such requirements have been successfully resisted by the applicants or by their universities in some instances but not in all.

Military and civilian security advisers to policy-makers are always concerned lest scientific knowledge should have technological applications that might benefit a potential enemy, and it is in their nature as professionals to opt for the safe course, that is to say, to put the onus on the scientist to show why the transfer of new knowledge (or, it now appears, even old knowledge) should not be restricted. It is not their business to consider whether the broader social costs of secrecy outweigh the supposed gains; but it should be the business of their political masters. Secrecy, though it may temporarily deny knowledge to another nation's scientists, also denies it to the scientific community at large within the nation in which it is generated.[30] The panel established by the National Academy of Sciences, referred to earlier, attempted to find a way through this dilemma. Its report recommended that no 'restriction limiting access or communication' should be placed on any area of university research, whether pure or applied, unless it related to a technology that met *all* of four criteria. The first was that the technology must be developing rapidly, with a short lead time from pure research to application; the second was that it should have immediately identifiable military uses. The applicability of these two criteria to a given area of research should be fairly readily decidable. The same cannot be said for the third criterion, which was that transfer of the technology would give the Soviet Union (no other country, even of the Soviet bloc, was mentioned) 'a significant near-term military benefit.' It is virtually impossible, in most instances, to demonstrate conclusively that a given piece of research will, if not kept secret, confer a significant benefit on another nation in the short run. Much will depend upon assessments of the capacity of that nation to make immediate or near-immedate use of the information. It is precisely this indeterminacy that leads the security-minded to play safe and opt for restriction. The fourth criterion poses an equally important question. It was that the United States should be the only source of information about the technology; or, if that were not so, that 'other friendly nations' that could also be the source had control systems at least as secure as those of the United States – or, in plainer English, that they would be prepared to co-operate in keeping the information to themselves. The likelihood of this happening is by no means evident.

In 1971, a long essay by Zhores Medvedev was published in the West after it had circulated in the Soviet Union as an underground 'self-publication' (*samizdat*). It was half-ironically entitled 'Fruitful Meetings between Scientists of the World' (Medvedev 1971). In it he argued that every branch of science today is a 'world science,' and it follows as a practical matter that 'a diverse and effective cooperation has become a necessary form of organisation of scientific work' (p. 117).[31] It is desirable at the very least to avoid unnecessary duplication of effort. For example, it is sensible for laboratories working in several countries on the elucidation of the structure of proteins to co-ordinate their efforts. The work is time-consuming and requires the use of expensive equipment, and it is absurd that effort and money should be expended by one laboratory on duplicating the work of another, especially since, as Medvedev put it, there are 'thousands of interesting proteins' to investigate. The amount of work being done throughout the scientific world in most areas of science is now so great that a scientist needs to have rapid access to the latest information in the field; and one of the best ways of meeting this need is by personal contact through conferences, seminars, and visits to research sites abroad.

The Soviet Union is notorious for the obstacles it has put in the way of such exchanges. Medvedev's essay is in part an account of his own frustrating experiences during the 1960s. One of the examples he quotes is the Third International Congress on Human Genetics, which took place in the United States in 1966. About one thousand scientists took part, but only four papers were submitted from the Soviet Union and none of the scientists who were to read them were allowed to attend. By contrast, many 'small science' countries were well represented: for example, sixteen geneticists from Sweden, twelve from Israel, and fourteen from Holland. It should be stressed, says Medvedev (pp. 133–5), that

> the chief means of critical appraisal of any piece of work, the demonstration of its strong and weak points, its methodological inadequacies and its position among other investigations in the same field is, even today, verbal, direct and immediate discussion in a circle of understanding colleagues. And this can be most useful ... when it is on an international basis ... We all know that in scientific work one successful thought that arises in some five-minute discussion, one critical remark, one piece of advice from a scientific colleague, can determine and change the course of work of a scientist for many years.

In a later work, *Soviet Science* (1978), Medvedev refers to the fact that the tight hold that party and government officials exercise over foreign travel by Soviet scientists is due to their 'primitive fear' of defection and of the 'ideological destabilisation' that may result from Western influences. Some increase in Western influence would certainly follow greater freedom to travel, but as Medvedev also points out, this influence is increasing anyway, especially among young people in the Soviet Union, a tendency quite unconnected with travel abroad. The restrictions imposed vary in severity according to the current state of international tension and especially of relations with the United States; but in any event foreign travel for Soviet scientists is a privilege and not a right. Indeed, it has become a weapon in the struggle against dissidence. If a scientist wishes to remain among those privileged to go abroad he must stay away from dissident colleagues, refuse to support them, and even, if need be, join in condemning them. It is too early to say what long-term effect the advent of *glasnost* may have on these problems.

For the Soviet Union, international freedom for scientists has been largely a matter of domestic policy; for the West, and especially for the United States, it is rapidly becoming a matter of foreign policy. Agreements aimed at encouraging exchanges between American and Soviet scientists (which frequently result from summit meetings between the two nations' leaders) have been abrogated or postponed for political reasons on several occasions; for example, by President Carter in 1979 in response to the Soviet invasion of Afghanistan. Scientists have become cynical about such high-level agreements, preferring where possible to negotiate their own informally. There is, indeed, something repugnant about resorting to politically arrived at agreements that, as John Ziman puts it, 'imply some commitment to peaceful transnationalism even between countries armed to the teeth with "scientific" weapons' (1978b). Nevertheless, scientists in the West have become embroiled in these political issues as a result of the Soviet treatment of dissident scientists. In an intended show of solidarity with their Russian colleagues many have refused individually to visit the Soviet Union; others have actively lobbied to have exchange agreements cancelled. But this issue has deeply divided the scientific community.[32] Hard-liners have argued that exchange programs, indeed all scientific co-operation, should be made conditional on the willingness of the Soviet authorities to cease their persecution of dissidents. Others have stressed the importance of keeping scientific exchanges open, arguing that severing all contacts leaves the dissidents more vulnerable than ever, feeling that they have been abandoned by the international scientific community. The U.S. National Academy of Sciences has itself, on a number of occasions, called for temporary boycotts.[33] In 1980, for example, its governing council imposed a six-month suspension of all seminars and workshops held jointly with the Soviet Academy of Sciences as a protest against the treatment of Andrei Sakharov and his rumoured expulsion (which was later confirmed) from the academy. In 1982, again in protest against the treatment of dissenters, it allowed to lapse an exchange agreement that had operated for many years.

As John Ziman has argued, the ideal of supranationalism in science is 'deeply rooted, both in the actually operative norms ... and in many famous historical precedents' (1978b). He calls it 'an enduring and fundamental principle on which the very concept of a republic of science is based.' The freedom of scientists to communicate and co-operate without hindrance from external authority is, for most scientists, an irreducible value (Shils 1984). But, the world being as it is, national security is also an irreducible value (which, again, most scientists share). These two values are not, however, wholly irreconcilable, and in pluralist societies with long traditions of freedom of speech and inquiry it should be possible to reconcile them without grave damage to either. The problem with the plea of 'national security,' as Sissela Bok has argued (1984), is that it is capable of indefinite extension. There are no limits in principle to the power of the state to declare the open communication of *any* scientific knowledge contrary to the interests of national security if 'national security' is sufficiently broadly defined. Yet the unrestricted use of this power would do irremediable harm to the progress of science, and it is scientific progress on which, paradoxically, the welfare and the security of nations now increasingly depend. It has sometimes been argued[34] that freedom of inquiry is not a basic right like freedom of speech. On the contrary, it is exactly like freedom of speech, with the same kinds of moral and practical dilemmas that face those who seek to exercise it.

SCIENCE AND THE MILITARY

This, then, is knowledge of the kind we are seeking, having a double use, military and philosophical.

Plato, *The Republic*

There have always been scientists who were willing to put their knowledge and skills at the disposal of their homeland when it was threatened by its enemies: for example, Archimedes (c. 287–212 BC) and, eighteen centuries later, Leonardo da Vinci (1452–1519), both of whom – as Bertrand Russell once remarked – were granted permission to add to human knowledge on condition that they also subtracted from human life. Archimedes' military innovations, which included improved machines for launching projectiles and a mirror for concentrating and directing the sun's rays against enemy ships in order to set them on fire, contributed greatly to the resistance of the citizens of Syracuse to the Roman armies in the Second Punic War.[35] Leonardo's notebooks are filled with details of his researches into weapons and fortifications, and one – the notebook on the nature, weight, and motion of water – mentions his design for an 'appliance for a submarine,' the specifications for which he refused to publish 'on account of the evil nature of men who would practise assassinations at the bottom of the sea' – an early instance of 'social responsibility' (a topic to be discussed briefly later). There was also Niccolò Tartaglia (c. 1500–77), a mathematician and translator into Italian of the works of Euclid and Archimedes, some of whose work was directed to the improvement of artillery fire. He, too, had misgivings. He wrote: 'One day, meditating to myself, it seemed to me that it was a thing blameworthy, shameful and barbarous, worthy of severe punishment before God and Man, to wish to bring to perfection an art dammageable to one's neighbour and destructive to the human race ... Consequently, not only did I altogether neglect the study of this matter and turned to others, but I even tore up and burnt everything which I had calculated and written on the subject.' But he changed his mind when his city-state was threatened by invasion by the Turks. Again, we find Galileo, in 1609, writing to his patron Leonardo Donato, the Doge of Venice: 'I have made a telescope, a thing for every maritime and terrestrial affair and an undertaking of inestimable worth. One is able to discover enemy sails and fleets at a greater distance than is customary ... Also on land one can look into the squares, buildings and defences of the enemy from some distant vantage point.' But these and other contributions to the practice of warfare were made by individuals long before the rise of 'organized science.'

The Great War of 1914–18 was the first in history in which the contributions of scientists were actively sought by the leaders of the warring nations, though in some instances tardily. In Britain, for example, an editorial in *Nature* in 1915 (vol. 95, p. 419) contained a cogent plea for a more effective use of scientists to survey 'the whole field of science to discover methods of destruction which we [that is, Britain] may use ourselves or from which our men may look to us [scientists] for protection.' Yet the innovations in war-fighting capability to which scientists contributed, like the tank, antisubmarine devices, and the airplane, came too late to materially affect the essential nature of the war, which remained a clash of massed armies on a gigantic scale – a soldiers' war rather than a scientists' war. The lesson was learnt, however, and scientists were used more systematically on military-related work in the period of uneasy peace from 1918 to 1939 – for example, in Britain, in the early development of radar. During the Second World War they were fully incorporated in the war effort.[36] An entirely new relationship was forged between them and those who directed and executed military policy. The first step was taken towards what Ziman has called 'the permanent mobilization' of science (1976: 330), a phenomenon that has been accompanied by a revolution in military technology in which scientists as well as engineers have played a large part.

World military expenditures now amount to about one trillion U.S. dollars a year. Roughly half of this total is spent by the United States and the Soviet Union; China, Britain, and France are among the other top ten spenders. A sizeable proportion of this outlay is devoted to research and development – in the United States in 1983 it was about 12 per cent of the total military budget, or some $40 billion – but the proportion varies greatly from country to country. In the United States in 1983 more than 65 per cent of the national (federal) budget for all R & D was for military R & D. In West Germany it was 15 per cent, and in Japan (a country whose constitution forbids it to go to war or maintain a standing army) it was 5 per cent. The nuclear powers devote a large part of their military R & D budget to developing new types of nuclear weapons and to improving the command, control, communications, and intelligence support systems needed for their deployment. But much military R & D money is also spent on inventing and improving 'smart' weapons, many of which have been made possible by rapid advances in computer science and in micro-electronics, in the production of which all the major military powers are now engaged to a greater or lesser extent. These weapons are revolutionizing the concept of conventional war, and many are now so lethal that they can be considered 'conventional' only in the limited sense that they are non-nuclear. Those under development include fuel-air explosives dispensing clouds of highly volatile fuel that, when ignited, can produce atmospheric overpressures capable of sinking an aircraft carrier even with a near miss, and missiles that can seek out targets as distant as two hundred kilometres and hit them with a destructive potential equivalent to three or four kilotons, or about one-third of the yield of the atomic bomb used on Hiroshima in 1945 (Rasmussen 1985).

Modern weapons are quickly obsolete, but as Harvey Sapolsky has aptly remarked, it is 'not rust or combat so much as the imaginations of the weapons designers that converts [them] into scrap' (1977: 451). It used to be said that necessity is the mother of invention. Now it seems – at least in the world of military technology – invention is the mother of necessity. Increasingly, scientists and engineers, not the military, prescribe what is needed – by demonstrating that new kinds of weapons are feasible and therefore should be made. A kind of technological determinism predominates in weapons policy, and military R & D has become self-propelling.

Two issues should concern us in the light of the themes developed earlier in this chapter. The first is how much pure (or basic) science – including pre-technological, mission-oriented pure science – is involved in this enormously expensive R & D effort; and the second is what effect military R & D has on the practice of basic research. There are, unfortunately, no definitive answers to either of these questions. Many books about 'science and warfare,' especially those written with a polemical intent, make statements such as the following: 'The military-industrial complex owns more scientists than any other sector of the economy' (Clarke 1971), or 'There are about a million scientists in the Soviet Union'; but these statements are usually based on international statistics that do not distinguish between scientists and engineers. The figures mostly provided are for 'QSE' – which stands for 'qualified scientists and engineers.' Other statistics that are used make no distinction between research and development, let alone between different types of science, although figures published by national granting agencies do usually attempt to give some breakdown for the latter.

A further problem, which is prone to mislead, is the fact that definitions of a 'scientist' vary from source to source. Derek de Solla Price once defined a scientist (legitimately enough for his stated purpose) as anyone who has published at least one paper in a recognized scientific journal; but 'recognized' obviously implies a value judgment and the image of a 'scientist' who has only published one paper is somewhat bizarre. Spencer Klaw in his book *The New Brahmins* defines a scientist as 'anyone who is eligible for membership in a recognized scientific body' (1968: 15). Using figures published by the U.S. National Science Foundation, 300,000 Americans met this test in 1966 – but two-thirds of them had only a bachelor's or a master's degree in science and 'a few thousand' had no degree at all. It is a recognized requirement for scientific research, in a university at least, that the researcher possess a doctorate. As Klaw nicely puts it: 'A scientist without a PhD is like a lay brother in a Cistercian monastery. Generally, he has to labour in the fields while the others sing in the choir.' In other words, not all those who possess a science degree are 'scientists,' at least in the sense used in this book. Nor is 'doing science' the same as being a scientist (laboratory technicians do science – sometimes better, I am told, than the scientists they work for).

Problems of this nature, statistical and definitional, make it possible for writers to get away with the most outrageously misleading statements. For example, Robin Clarke states, under a subheading 'Colonizing Civilian Science,' that 'the universities ... have become the outstations for defence departments' (1971: 182) – a remark that takes no account of the distinctions between scientists and engineers and between research and development. He later quotes, with approval, a statement by Benjamin Farrington (in a book about Greek science) that 'in the space of 300 years or a little more, the face of the world has been transformed. But so has the image of the scientist. Research, now multiplied a millionfold, is mainly for war' (p. 189). This is patently absurd. Fortunately we have at least some orders of magnitude. First, as already stated, much of the money devoted to research and development by the governments of the industrialized countries is spent on military R & D (though the proportion varies greatly from country to country – as measured, for example, by per capita expenditures). Secondly, most of this military R & D expenditure is spent on development, not on research. Thirdly, such figures as we have make it clear that the amount of *basic research* that is done in direct support of military ends is, proportionately, quite small. For example, in 1979, 80 per cent of the U.S. Department of Defense budget for R & D was allocated for development, and less than 3 per cent was allocated to basic military research. More than 90 per cent of *total* U.S. government investment in all types of basic research was devoted to research not directly related to military use. While these percentages represent considerable sums of money (the United States being one of the largest military research spenders) they give the lie to the contention that scientific research is nowadays done mainly for war.

The precise effects of burgeoning military research on the present-day practice of pure science are hard to assess, and this problem has been little studied. It is obvious, however, that military science creams off much scientific talent that would otherwise be available for 'peaceful' research, since the pool of really talented people is limited. Working for a defence agency has undoubted attractions. One, especially for younger and unestablished scientists, is the shortage of positions in the universities. As one young scientist doing weapons research for the Strategic Defense Initiative at the Livermore National Laboratory in California told William Broad, 'it was either this or a beet factory' (Broad 1985). Military research is better funded, and military research establishments usually provide the latest 'state of the art' equipment and excellent laboratory facilities. A scientist working in a defence establishment (as opposed to one who works in a university on a defence contract) is sheltered from the competition that characterizes the scientific world outside, has no worries about the struggle for tenure, is free of personal involvement in fund-raising, and has the problems on which he works set for him[37] (Sutton 1984).

Expanding military budgets for research and development in practice usually entail less money for civil R & D and hence less for basic research. Critics of the administration of military research argue that projects are frequently overfunded, and that assessments of their merits before they are authorized are lax and subsequent quality controls inadequate. There is much evidence that this is often true. As Sapolsky observed, scientists and engineers who are proponents of particular weapons systems 'tend to exaggerate the military benefits that are likely to accrue ... and to depreciate the technological and political risks' (1977: 452). These are very real. Military research today is peculiarly vulnerable to changes in domestic policies and the state of international relations. While it would be unwise to claim that money spent on basic, 'peaceful' science is never wasted, it is certainly true that money spent on 'warlike' research is often wasted – not only because the project has been overfunded or the money misapplied by those doing it, but also because the end product is either obsolete by the time the R & D cycle is completed or changes in government policy have rendered it redundant.

Most military-related basic research is classified, although university scientists and administrators challenge this wherever possible (see previous section). Many university scientists throughout Europe and North America have refused to solicit or to accept money for research supporting the U.S. Strategic Defense Initiative. Many, no doubt, have done so because they believe that the program will prove to be ineffective as a defensive shield against nuclear attack and strategically unwise in that it is likely to lead to an escalation of the nuclear arms race; but it is probable that others have done so simply because they object to security provisions that would forbid publication of their results and restrict normal scientific intercourse. The effects of secrecy are most onerous, however, when research is done within a closed institution such as a chemical and biological warfare laboratory like Porton Down in Britain, Fort Detrick in the United States, and Suffield and Shirley Bay in Canada, or in nuclear-weapons research establishments such as Los Alamos, the Centre d'Etudes Nucléaires at Grenoble in France, and Aldermaston in Britain.

For scientists engaged in secret work there can be no general peer review and no, or only very restricted, publication of results. In a closed institution it is likely that there will also be little participation in professional societies (Sutton 1984).[38] In his study of scientists at the Livermore National Laboratory in California (which does much basic, pre-technological science not related to weapons as well as applied research on nuclear weapons – see note 38) Sutton found that communication between researchers *even within* the laboratory was highly compartmentalized according to the doctrine of 'need to know,'[39] although there was a certain amount of 'leaking' by scientists who were veterans of the laboratory and had moved over the years from project to project, and by younger scientists (often graduate assistants) who had less respect for the system. Sutton reported that 'scientists at Livermore operate outside the mainstream of academic science,' adding 'more so in weapons research than in non-weapons research' (p. 210), which implies that scientists engaged in basic research at Livermore not directly related to weapons (for example, research on fusion for nuclear power) were affected by the fact that it was done in a primarily military environment (the entire laboratory, which covers many acres, is enclosed in a secure perimeter). Indeed, he states that participation in 'major professional activities' was not notably higher among those not doing weapons-related research than it was among those who were; and that for the former the fact that access to all parts of the laboratory required security clearance made outside contacts such as those with foreign scientists and consultants from American universities difficult. One informant told him: 'It's hard to bring foreigners in here, which bugs me, and bugs just about everybody because science is pretty international and that's one of the beauties of it' (p. 212). Further difficulties for those engaged in basic science were created by the fact that some research (for example, laser research) is partly classified and partly unclassified.

The role of science and technology in modern war is, as Ziman has said, 'all too evident, and much of the abhorrence of war amongst civilised people has been transferred to science itself' (1976: 302). But to condemn science unequivocally is as unfair as it is illogical. Today's scientists are faced with exactly the same dilemma as faced Leonardo da Vinci and Niccolò Tartaglia centuries ago. The universality of science 'is completely antithetic to militarism and aggressive nationalism' (Ziman 1976: 335), but these are facts about the world in which we live. Even the United Nations Charter acknowledges the right of nations to go to war in their own defence (Article 51). Some scientists in recent times have, indeed, refused to work on projects that are in any way related to military use – which, in a free society, is their right[40] – but often, discreditably, with unfortunate consequences for their careers. An example is Steven Heims, the author of a book about John von Neumann and Norbert Wiener (1980), mathematicians who played a crucial role in nuclear weapons research. Heims withdrew from military-related research in the 1960s, and much of his book was written during periods of unemployment that followed this decision and between temporary teaching posts. Wiener himself took this step in 1946, writing, 'I do not intend to publish any future work of mine which may do damage in the hands of irresponsible militarists' (Heims: 208); but Wiener was internationally known and an established figure in the scientific world with a vast number of published papers and had much less to lose than a younger man.

Ziman comments that 'the sad thing that has to be said is that most scientists employed in military research are quite complacent about it. They ... do their technical jobs with complete devotion'; but, he adds, 'this is not a criticism of their moral judgement, it is simply a statement of fact' (1976: 334). There are, however, exceptions – scientists who no doubt work 'with complete devotion', but nevertheless with considerable inner disquiet. An example, taken from Broad (1985), is Peter Hagelstein. Hagelstein was responsible in 1979 at the age of twenty-four, at the Livermore National Laboratory, for solving the difficult theoretical problems involved in the construction of a nuclear-pumped X-ray laser, an improved version of which has since become one of the centrepieces of the Strategic Defense Initiative. His hope was to win a Nobel Prize for a laser that would have no military use but would be of immense benefit in biology and medicine. Now, as Broad says, he appears to exhibit a deep ambivalence towards his brainchild. 'My view of weapons has changed,' he is quoted as saying. 'Until 1980 or so I didn't want anything to do with nuclear anything. Back in those days I thought there was something fundamentally evil about weapons. Now I see it as an interesting physics problem.' There is more than an echo here of Robert

Oppenheimer's famous dilemma. Initially opposed to work on a hydrogen bomb on moral grounds, he changed his mind when he became convinced that it was, in his words, 'technically sweet.'

Individual conscientious objection to military research is tolerated in free societies, but so is collective action aimed at persuading governments to change their policies. Scientists have played a leading and often crucial role in this process since the end of the Second World War: opposing exclusive military control over the production of nuclear weapons (in 1945), leading a campaign against the deployment of an anti–ballistic missile system, and supporting the Nuclear Test Ban Treaty of 1963.[41] The most recent political actions by scientists have been in opposition to the Strategic Defense Initiative and to a resumption of the production of biological weapons. Many of the leaders in these campaigns have themselves made major contributions at one time or another to weapons research. A host of quasi-permanent organizations are involved: for example, in the United States, the Union of Concerned Scientists, which, among its many other activities, in 1983 sponsored a 'Call for a Halt to the Arms Race' by 1500 physicists, including over twenty Nobel laureates; in Canada, an organization called Science for Peace; and similar organizations in many other countries. There have also been numerous ad hoc protests such as participation in opposition to SDI from scientists in North America, Britain, and other Western European countries. Somewhat surprisingly, there have been fewer public protests by scientists about the proliferation of, and flourishing international trade in, high-technology conventional weapons, much of which is conducted with often unstable regimes in Third World countries. Indeed some scientists, among them Freeman Dyson of the Institute for Advanced Study in Princeton, New Jersey, have argued that the increasing deadliness of 'smart' weapons will reduce the likelihood of nuclear war since similar destructive effects will eventually be obtained without the radiation and other hazards associated with nuclear devices.

4

Perversions of Science

The human understanding is no dry light but receives an infusion from the will and affections; whence proceed sciences which may be called 'sciences as one would.'

Francis Bacon, *The New Organon,* Aphorism 49

One of Bertrand Russell's popular (or, as he characteristically called them, unpopular) essays is entitled 'An Outline of Intellectual Rubbish' (1950). In it he poked fun at the follies of theologians, prophets, cranks, and Aristotle, but said nothing about scientists. Yet the history of science is by no means devoid of evidence of ill-founded, prejudiced, silly, and even vicious ideas. It is one of the strengths of the (historically) collective and corrective nature of science, however, that rubbish is eventually recognized and exposed; but sometimes, alas, not before it has done much harm. It makes sense, therefore, to speak of instances of the 'perversion' of science, not merely by non-scientists (which instances are all too frequent), but by scientists themselves; and I use the term 'perversion' here in several of its dictionary meanings – to misconstrue, to misapply, to lead astray, and to turn aside from its proper use. There are those (including many contemporary sociological critics of science) who would argue that the concept of 'perverted' science is illegitimate since the science of any given historical period is shaped by the social conditions, ruling ideologies, and productive forces then prevailing – in other words, that science necessarily reflects the prejudices, beliefs, and opinions of the age in which it is done. As should be apparent in many places in this book, I do not share this charitable (or uncharitable) view. Scientists do reflect in their work – to some degree – the visions of their age, and they are influenced by the opinions and beliefs of their fellow beings. It could hardly be otherwise. But it does not follow that they must be apologists for the established order, or that they merely mirror the social context. Stephen Jay Gould (1981a: 22) has very sensibly argued that although science is 'a socially embedded activity' and 'culture influences what scientists see and how they see it,' there is no need to abandon the idea that an autonomous, 'objective' reality exists and that science, 'though often in an obtuse and erratic manner,' can learn about it. Science, though 'embedded in surrounding culture,' can nevertheless be 'a powerful agent for questioning and even overturning the assumptions that nurture it.' As we shall see from the examples discussed in this chapter, science is frequently 'perverted' for want of this critical faculty in those doing it, because of a lack of appreciation of its own proper method, adherence to an outdated metaphysics, or a dogmatic clinging to ideas that rest on little or no evidence.

Not surprisingly, the most striking instances (though by no means the only instances) of the perversion of science have been associated with what we now call the 'life sciences,' since these seem to us to impinge most directly on the human condition and to have the profoundest political, moral, and social consequences. While we can be dispassionate about the fate of the solar system a billion years hence, we cannot be about the nature of humankind here and now. The Oracle's injunction, Know Thyself, has never been a comfortable doctrine. It is in this domain, then, that the greatest temptations exist to distort the truth, where we are most prone to be blinkered by our preconceptions, where values extraneous to science are most likely to intrude, and where knowledge is most likely to be interpreted to legitimize and maintain the interests and the power of dominant social groups. Yet it is precisely in this domain, and for these reasons, that we should be most critical, most aware of our assumptions, and most cautious in our conclusions.

In the article 'Human Nature' in the *Encyclopaedia of the Social Sciences* (1937), John Dewey argued that the significance of the idea was 'gathered about three questions.' The first was whether the nature of human nature was such that certain social arrangements were likely to be successful while others were necessarily doomed to fail. Was war, for example, inevitable? Secondly, how far was human nature 'modifiable by deliberate effort'? How were heredity and environment related to one another? The third question was how great and how fixed was the range of variations in human nature between individuals and between groups. 'Are some racial and social groups by nature definitely inferior to others because of causes which cannot be altered?' And the same question, he said, could be asked about individuals within each group. These questions, Dewey said, are the source of 'controversies involving intense feeling ... It is therefore extremely difficult to attain impartiality with respect to them, and discussions are often apologetics for some position already assumed on partisan grounds.' The most recent examples of such controversies form the main subject-matter of this chapter: the renewed debate about racial and group differences sparked by the publication in 1969 of an article by the American educational psychologist Arthur Jensen (the 'IQ controversy'), and the controversy over the alleged 'new science' of sociobiology associated chiefly with the name of Edward Wilson. Before discussing these cases in detail, however, it is necessary to look briefly at the nineteenth- and early-twentieth-century background to which Dewey was indirectly referring. For clarity of exposition I shall alter slightly the order of his questions, dealing first with racial and group differences, secondly with the social aspect of Man's 'animal nature,' and lastly with the question of the modifiability of that nature by deliberate effort.

THE HISTORICAL BACKGROUND

Race and Group Differences

Of all the vulgar modes of escaping from the consideration of the effect of social and moral influences on the human mind, the most vulgar is that of attributing the diversities of conduct and character to inherent natural differences.

John Stuart Mill, *Principles of Political Economy*, Book I

The concept of race, and especially the idea of inherent racial differences, is of relatively recent origin. In a highly diffuse form (in-group and out-group, the Chosen People, Greek and Barbarian) it is deeply rooted in antiquity; but until the great voyages of discovery were undertaken by Europeans beginning in the fifteenth century, and until subsequent exploration extended the known world and brought Europeans into contact with representatives of its varied peoples, 'race' in a specific sense could have little meaning. As further expeditions of venture and trade and, later, colonization took place, ethnographic observations accumulated and a powerful stimulus was provided for generalization. 'Race,' as a term for a group of living things, seems to have made its appearance in English in the late sixteenth and early seventeenth centuries; but its first use in relation to Man tended to encompass the whole of humanity ('the human race'). This is how Shakespeare and Milton used it.[1]

The truth of the matter is that the very concept of race, let alone the issue of specifically *racial* differences, both of which bulked large in nineteenth-century thought and action, is nowadays regarded by biologists as highly questionable. All human beings belong to the same natural species and all human groups are mutually fertile.[2] This idea is not new. It was advanced by Georges-Louis Buffon (1707–88), the great French naturalist, but its significance was little noted. The line of ascent of *Homo sapiens* from early Man, and from our pre-human ancestors, still lacks the necessary fine detail, but the general pattern of animal evolution suggests that the distinctive physical differences that came to exist between the major groupings of humankind (Eurasian, Mongoloid, and so on) were the result of differentiation from a common stock through adaptation and natural selection within specific ecological niches. Biologically, then, there is very little to be said for race. The basic mistake made by many of our nineteenth-century forbears, who lacked the knowledge (however imperfect) that we now possess, was in thinking that the major groupings of humankind (conceived in terms of their distinctive, visible, and mainly physical characteristics) constituted separate species. This is not to say that 'races' have no existence; but they exist only as socio-cultural entities, not as biologically meaningful aggregations. Humankind *is* divided by race – not in a scientifically objective sense, however, but by virtue of the twin assumptions that races exist 'in nature' and that racial differences are important. 'Race' is a social image, not a scientific construct. Yet the entire history of thinking about race since the eighteenth century is marked by a succession of attempts to provide scientific reasons for believing these assumptions to be true.[3] Along the way, other human differences have been asserted to be inherent in nature: principally those of social class and between male and female.

The taxonomic division of humankind into separate races began with the first great modern taxonomist, Carolus Linnaeus (1707–78); the attempt to provide a scientific basis for racial differences with the German comparative anatomist Johann Blumenbach (1752–1840). The method adopted was the careful measurement of physical characteristics (anthropometry) and especially the measurement of the skull (craniometry); and the importance of racial differences was then reinforced (though this did not follow of necessity) by ordering the measured characteristics into a hierarchy of superior and inferior types.

The earliest craniometrists gave much attention to the facial angle of the skull. This had been thought by Aristotle to mark degrees of intelligence; and the Greek sculptors' 'ideal' facial angle, which they used when representing the beauty of the gods, was taken as the norm. Blumenbach, however, seems to have been sceptical about 'beauty' as a differentiating criterion, remarking that he supposed that even a toad looked beautiful to other toads. Measurement of the volume of the skull (cranial capacity) was soon added to the study of its shape, and this led to debate about the relative merits of lead shot, mustard seed, barley, mercury, sand, and other materials used for doing this – and with good reasons since the manner in which the material could be packed into a skull by an investigator could affect the result. Later, brains removed from cadavers were measured by weighing. All this presupposed that brain size was related to intellect and other qualities: the larger the brain, the further developed on the evolutionary scale the race it represented was assumed to be. When this presupposition ran into difficulties, as for example when it was discovered that the Negro brain was larger than that of the Chinese and the 'Hindostanee,' attention shifted to an examination of brain structure.

The craniometrists were serious-minded and devoted scientists. They included such luminaries as Paul Broca in Europe and Samuel Morton in North America. Broca (1824–80) was a distinguished surgeon and anatomist with many accomplishments.[4] He is said to have made more than 180,000 skull measurements during his working life and he devised many ingenious instruments for use in craniometry and developed a 'craniological index' of the proportions of the skull. Samuel George Morton (1799–1851) was a Philadelphia physician who 'amassed the world's largest pre-Darwinian collection of human skulls' (Gould 1978: 503). His collection was known affectionately as 'the American Golgotha.' Like all his fellow measurers, he 'gathered skulls neither for the dilettante's motive of abstract interest nor the taxonomist's zeal for complete representation' (Gould 1981a: 51). He was bent on proving a point: that the measurement of the physical characteristics of the brain could objectively establish a ranking of the races of humankind.

The passion for measurement and for the 'hard data' that it is supposed to yield is a positivistic thread running through much in nineteenth-century science and it lingered well into the twentieth-century. It could be said to have reached its finest flowering in the philosophy of science of Karl Pearson (1857–1936). Pearson was trained as a mathematician and became one of the founders of modern statistical method.

Since he was also a Social Darwinist and a supporter of eugenics, we shall meet him again in the two following sections of this chapter. His disastrous philosophy of science is contained in a once influential book, *The Grammar of Science*, which he published in 1892. In it he argued that science should take note only of those things that can be observed and, preferably, measured; that the aim of science is description, not explanation; and that its function is to classify facts and 'recognise their sequence and relative significance.' Though he was immediately influenced by the phenomenalism of the German physicist Ernst Mach (1838–1916), he was firmly within the tradition of the craniometrists and shared their belief in the objectivity of numbers. In this belief both he, and they, were sadly misled. The idea that a number produced by a measurement is of necessity objective – a 'hard fact' – is (to us) obviously fallacious since the measurement may itself be affected by subjective factors: desires, prejudices, biases, which may often be unconscious. Pearson seems to have been wholly unaware of this possibility. 'There is only one solution,' he wrote, 'to a problem of this kind' (the problem of the immigration of Jews from Eastern Europe), 'and it lies in the cold light of statistical inquiry ... We have no axes to grind, we have no governing body to propitiate by well-advertised discoveries; we are paid by nobody to reach results of a given bias ...' 'We firmly believe,' he continued smugly, 'that we have no political, no religious, and no social prejudices ... We rejoice in numbers and figures for their own sake and, subject to human fallibility, collect our data – as all scientists must do – to find out the truth that is in them' (quoted by Gould 1983: 296). It is hardly plausible that Pearson, one of the founding fathers of biometrics, should have believed that numbers have the capacity to speak to us without prior interpretation.

In his splendid book *The Mismeasure of Man* (1981a), Stephen Jay Gould has rendered a signal service by actually reworking much of the data used by Morton and Broca, and has shown conclusively a number of important things. First, that many of their conclusions were based on inadequate data (the data were insufficient to bear the weight put upon them); secondly, that, even where adequate, the data were frequently misrepresented or manipulated – often unconsciously but sometimes deliberately – in order to substantiate the conclusions; and thirdly, that data was sometimes fabricated. Further, as Gould points out, prior prejudices may *prevent* adequate data from being assembled – and there are indications of this occurrence also. In sum, craniometrists like Broca and Morton were guilty of using numbers to illustrate prior conclusions, even though they were sometimes unaware that this was what they were doing.

The conclusions that they wished to establish – which were the assumptions from which they began – were, generally, that some 'races' (and principally the white 'race') were superior to others; and that men were in certain respects, for example in intelligence, superior to women. But these were simply among the prejudices of their age.

Morton's work on a great diversity of human skulls (including more than a hundred from the tombs of Ancient Egypt) was his monument; and the tables summarizing his results, arranged by race, constituted his chief claim to fame. Gould says: 'They represent the major contribution of American polygeny to debates about racial ranking. They outlived the theory of separate creations [polygeny] and were reprinted repeatedly during the 19th century as irrefutable, "hard" data on the mental worth of human races' (p. 53).[5] Since Morton published all his raw data and explained in detail their origin, Gould could infer with a high degree of accuracy 'how he moved from raw measurements to summary tables,' and thus was able to show that these tables are 'a patchwork of fudging and finagling in the clear interest of controlling [sic] a priori convictions.' Yet he could find no evidence of conscious manipulation, and comments that had Morton been a 'conscious fudger' he would not have published his data so openly. The evidence he did find includes improper guesswork (of the age and sex of the specimens, for example); errors in computation that usually support the conclusions Morton wishes to reach; shifting criteria (for example, including Inca Peruvians to decrease the Amerindian average, but deleting Hindus to raise the Caucasion mean); sampling errors; and the indiscriminate mixing of results obtained by different measuring techniques (for example, mustard seed and lead shot). Having reworked Morton's data, and made necessary corrections, Gould concluded that they revealed no significant differences in cranial capacity between the races (the values range from 87 cubic inches for the Mongolians and modern Caucasians to 83 cubic inches for the Africans). (All this is detailed in Gould 1978 and in 1981a: 50–69.) As an example of the weight that craniometrists were apt to place on the slenderest of evidence, we should note what Morton said of the Shoshone Indians.[6] 'Heads of such small capacity and ill-balanced proportions,' he wrote in 1848, 'could only have belonged to savages ... A head that is defective in all its proportions must be almost inevitably associated with low and brutal propensities, and corresponding degradation of mind.' This definitive pronouncement was based on his examination and measurement of three skulls!

Gould's studies of Broca reveal other interesting aspects of what might be called 'the craniological legend.' Crucial to this legend is the equating of brain size with intellect and other qualities while ignoring the fact that, if any conclusions at all are to be drawn from brain size, adjustments must be made for body size. There are small clever people with impeccable morals and there are large stupid people who are libertines. The fact that, for example, women's brains are on average smaller than men's is related to the fact that women are on average physically smaller than men. It is the ratio of brain size to body size that is important. Broca recognized this fact. Morton seems to have been unaware of it; it does not enter into any of his calculations; but Broca wished to ignore it (in at least this instance) because he was convinced that women could not equal men in intelligence. His data, obtained from some 450 autopsies on males and females in Paris hospitals, showed a 14 per cent difference in average brain weight between the sexes. When challenged on the issue of body size, he was led to produce the following extraordinary example of circular reasoning: 'We might ask if the small size of the female brain depends exclusively upon the small size of her body ... But we must not forget that women are, on the average, a little less intelligent than men, a difference which we should not exaggerate but which is nonetheless real.' Broca was a careful and competent investigator. Gould found his numbers sound. It is his inferences that are at fault. (Gould 1980a, chapter 14).[7]

Broca was also convinced, as were so many of his contemporaries, of the possibility of objectively ranking races. To try to establish this ranking more firmly he experimented with a variety of anthropometric measures other than cranial capacity. Gould chronicles his struggles with these, and also his near-abandonment of brain size when the evidence led to implications that he did not want to accept – for example, that if brain size were the only criterion, 'Eskimos, Lapps, Malays, Tartars, and several other peoples of the Mongolian type would surpass the most civilised people of Europe' (this was written in 1873, seven years before his death at the early age of 56). He appears to us as an honest man struggling to explain away inconvenient facts; and as is true of many honest intellectuals with their backs to the wall, his wrigglings can best be described as concessions without the admission of defeat.[8] For example, he admits that 'a lowly race may have a big brain' (the same source, 1873) and that therefore the volume of the brain cannot play 'a decisive role' in the ranking of races; nevertheless it has 'a very real importance' – a statement that he attempts to justify by the argument that, though big brains *may* be found among inferior races, small brains are found *only* there.

The belief in ranking, the belief that human variations are naturally linear and hierarchical, forced the craniometrists into error. The belief in the objectivity of numbers, important though it was in determining method, was secondary. Numbers were used to give strength to the prior belief. The compulsion to rank is captured in a statement by Louis Agassiz (1807–73), one of the most respected and brilliant of nineteenth-century zoologists: 'There are upon earth different races of men, inhabiting different parts of its surface ... and this fact presses upon us [scientists] the obligation to settle the relative rank among these races.'[9] It is far from obvious why this obligation should have been thought to exist, but two factors may help to explain it. The first is a strongly held belief in the fixity of natural characteristics that was embodied in the highly influential Linnaean scheme of taxonomy. Linnaeus, deeply imbued with the Aristotelian metaphysics of Aquinas, was an essentialist; that is to say, he held that every living thing was characterized by certain features (which it was the business of the taxonomist to uncover) that constituted the essence, the essential nature, of the thing. Thus all the members of a species, for example, shared a certain fixed essence, however they might differ individually in specific, and therefore only contingent, ways. The biologist Ernst Mayr has argued (1982) that this essentialist view of nature constitutes one of the two great world-views that have underlain the entire history of biology. The other view, the 'population' view, holds that it is not the possession of a common essence that makes a collection of individuals a species but rather their potential to interbreed. This, as we have seen, is the modern view. Although it was held by some early modern naturalists like Buffon,[10] this view could not gain a real foothold before the acceptance of Darwinian evolutionary theory (which was not fully achieved during the nineteenth century), or ascendancy before the rise of modern genetics.

The essentialist view of nature implied that, if humankind were indeed divisible into separate species (or even into subspecies of the species Man), that is, into distinct 'races,' then each would possess innate features that would differentiate it essentially and not merely contingently from the others. This by itself would not necessitate ranking in terms of natural superiority-inferiority, but taken together with a second factor it would. This was the medieval belief, underlying Linnaean taxonomy, and still widely though no longer universally accepted in the first half of the nineteenth century, that the whole of creation was fixed and ordered hierarchically. 'All things whatsoever observe a mutual order,' Dante had written in the *Divine Comedy*; and four centuries later Alexander Pope still proclaimed the 'vast chain of being, which God began.'[11] Although by the eighteenth century the notion of a Great Chain of Being had been largely de-Christianized, it remained important as a secular world-view.

The history of craniometry and anthropometry in the nineteenth century affords an excellent example of the search for scientific reasons for believing in the existence of 'natural,' innate, essential differences between human groups. It also shows how easy it is for a faulty methodology to take hold and dominate the thinking even of persons who are, or who sincerely believe themselves to be, committed to the search for truth. It demonstrates, further, in a very plain way, how preconceived ideas can lead to self-deception (even mass self-deception) in science, and thus to the committing of egregious errors. Craniometric arguments, Gould says, 'lost most of their luster in our century as determinists switched their allegiance to intelligence testing – a more "direct" path to the same invalid goal of ranking groups by mental worth – and as scientists exposed the prejudiced nonsense that dominated most literature on form and size of the head' (1981a: 108). He adds, in a striking phrase, 'what craniometry was for the 19th century, intelligence testing has become for the 20th' (p. 25). As we shall see later in this chapter ('The IQ Controversy'), there is the same obsession with measurement and 'hard data'; the same belief that measurement (or, more sophisticatedly now, what one does with measurements statistically) is objective; the same compulsion to rank whole groups of human beings according to fixed characteristics.

The Social Consequences of Man's 'Animal Nature': Social Darwinism

And as every great scientific conception tends to advance its boundaries and to be of use in solving problems not thought of when it was started, so here, what was put forward for mere animal history may, with change of form but an identical essence, be applied to human history.

Walter Bagehot, *Physics and Politics*, 1872

Social Darwinism, one of the great intellectual disasters of the nineteenth century, was the extension of misunderstood Darwinian evolutionary theory to human races, classes, and nations. The notion that a 'law of evolution' applied to human history (historicism) long pre-dated the publication of the *Origin of Species*, but Darwin's ideas were eagerly seized upon to explain how that evolution worked. Darwin's evolutionary theory is not, however, a law of change but an explanation of *how* change takes place in nature *when* it takes place. This was his great contribution to human thought: he provided the first detailed and empirically supported explication of the mechanism of the evolution of living things. As the first part of his title suggests, it is a theory about the *origin* of species; that is, how new living forms arise by the operation of natural selection. It is only in that limited sense a historical theory. It does not predict, or give any indication of, *the future course of change* – which Social Darwinists mistakenly took it to do. Natural selection, as Ghiselin has pointed out (1969: 66), does not even imply that evolution will occur; it only implies that when evolution does occur it will proceed according to Darwinian rules. Still less is it a law of progress. There is nothing about 'progress' in the early editions of the *Origin*. Indeed Darwin deliberately sought to avoid the use of the term 'evolution' precisely because of its common association with the notion of progress, preferring the more neutral formulation 'descent by modification.' And in so far as Darwin may have toyed with the idea of progress in some of his later work, and especially in the *Descent of Man* (1871), this was no doubt due in part to the influence of his age,[12] in which the idea of progress was a prominent theme, and possibly also to the fact that he did not fully understand his own theory. (This would be in no way remarkable, since many of the great scientific theories have not been fully understood by their originators or by those who first accepted them.)

Still less again is Darwinian evolutionary theory a law of progression from the inherently 'less fit' to the inherently 'more fit' (as, again, many Social Darwinists tended to assume it was). That was not what Darwin intended by the phrase 'survival of the fittest.' The wings of a certain species of moth found in Britain turned black during the nineteenth century, at least in heavily polluted areas of the industrial north and midlands; they were the 'fittest' to survive because tree trunks on which they rested had been blackened. With the coming of the Clean Air Act in 1956 they became 'less fit' and the proportion of the moth population with black wings fell dramatically. Natural selection is a motive 'force' for Darwinian evolution in only a metaphorical sense. It is not a force acting on nature. It is, as Ghiselin says, a description: a description of what happens, when certain conditions are met, to the relationships between the component units of a population. Many people have held that the expression 'the survival of the fittest' is a tautology, but this interpretation has only arisen because natural selection has been improperly understood.[13] Nor, especially, does it mean that 'being the fittest' will guarantee survival, an idea that underlies doctrines of racial purity. Being 'fitter' applies to individuals within a population, not to whole populations. Neither, lastly, is there anything in Darwinian theory about 'higher' and 'lower.' Darwin did not postulate a hierarchical order of nature, and indeed went out of his way to deny it. 'Higher' and 'lower,' he held, have no meaning in nature; he warned specifically against this anthropomorphic fallacy. Yet all the variants of Social Darwinism were based on one or more of these mistakes.

Social Darwinist doctrines did not, of course, go unchallenged, but they were extremely influential in the last three decades of the nineteenth and the early years of the twentieth century, and frequently had an important impact on thinking about public policy. In America, Social Darwinism was used to justify the existing social order and to support an ideology of aggressive industrial capitalism and the continued dominance of the Anglo-Saxon 'race'; in Germany it was used to legitimize policies of national aggrandizement and expansionism and to bolster the doctrine of the innateness of human differences and, hence, ideas of racial 'purity'; in Britain it was taken up less enthusiastically but provided, nevertheless, a strong underpinning for eugenics and for 'meritocratic' educational policies. Darwin's 'abominable volume,' as he called it, no doubt gave unintended support for these tragic misunderstandings, for he subtitled it 'The Preservation of Favoured Races in the Struggle for Life'; and the concepts 'survival of the fittest' (which Spencer, not Darwin, coined but Darwin later inserted in his work) and 'struggle for existence' were obviously open to misinterpretation. I shall take as two important scientist-exemplars of Social Darwinism Ernst Haeckel in Germany and Karl Pearson in Britain. In America the major exponents tended to be social rather than natural scientists, one of the most influential being the forbiddingly puritanical professor of political and social science at Yale, William Graham Sumner.

Ernst Haeckel (1834–1919) was a widely respected scientist in his native Germany and, as professor of zoology at the University of Jena (to which post he was appointed in 1862 and where he remained until the end of his long life), he became an immensely popular teacher and public lecturer and attracted many pupils who later became famous themselves.[14] By the end of the century he had become, for a wide circle of devotees, 'the wise old genius of Jena.' It was he who popularized the name of Darwin in Germany, but only at the cost of grossly misusing Darwin's ideas. Philosophically, Haeckel shared with many of his German contemporaries a deep respect for the mystical romanticism of *Natur-*

philosophie (see chapter 6). In his hands, Darwinism became much more than just a powerful scientific theory; evolution was elevated to the status of a cosmological principle, the unifying world-force that nature philosophers had long sought. Evolution, he declared in 1866, 'is henceforth the magic word by which we shall solve all the riddles that surround us, or at least be set on the road to their solution.'[15]

The theory of descent by modification was, in fact, only a small part of Haeckel's grand design, and Haeckel greatly overstated his debt to Darwin. The difference between the two men is clear in their attitude to the study of nature. Haeckel was convinced that 'a vast, uniform, uninterrupted and eternal process of development obtains throughout nature' from the most primitive organism to Man himself (he called his system Monism) and, as a consequence, that all the facts of nature would be found to conform to this view. Darwin made no such assumption. Thus, for example, on the question of the origin of life Haeckel argued that if the hypothesis of spontaneous generation (one of the great themes dividing nineteenth-century biologists) were not accepted it would be impossible to 'proclaim the unity of all nature, and the unity of her laws of development.' Darwin on the other hand wrote that he 'long regretted' his 'truckling to public opinion' on the matter of an original divine creation of life, when what he had really meant to say was that the process of creation was 'wholly unknown.'[16] Spontaneous generation was at that time 'beyond the confines of science'; it was 'mere rubbish, thinking at present of the origin of life; one might as well think of the origin of matter.' And he saw no reason why it was necessary to show how *life* had arisen before one could show how the *forms* of life arise. 'This seems to me about as logical ... as to say it was no use in Newton showing the law of attraction of gravity and the consequent movement of the planets because he could not show what ... gravity is.'[17]

The popular but unfounded belief that Haeckel was the German champion of Darwinian thinking is paralleled by an equally potent myth concerning his political ideas: the myth that Haeckel was a liberal, progressive free thinker. This notion has recently been repeated in Ghiselin's otherwise excellent book *The Triumph of the Darwinian Method.*[18] Ghiselin argues that Haeckel has been 'persecuted' because he was a popular leader in the nineteenth-century struggle against Prussian despotism and that his present bad reputation is largely due to the fact that people have believed what 'reactionaries' have said about him. Nothing could be further from the truth, as Gasman (1971) has conclusively shown in his book on the scientific origins of national socialism. Haeckel was a believer in strong, authoritarian rule, and a devoted follower of Bismarck, who adopted a strident pan-German nationalism. His Social Darwinism took an extreme and almost gloatingly pessimistic form, for he believed not merely that the survival of the fittest in its most literal and naive sense applied to human groups, but also that natural selection 'teaches us that in human life, exactly as in animal and plant life, at each place and time only the small privileged minority can continue to exist and flourish; the great mass must starve and more or less prematurely perish in misery.' This belief formed one of the bases for his advocacy of a ruthless program to eliminate the 'unfit.'[19]

The contrary view to Haeckel's gloomy and dogmatic biological determinism, argued by men like Thomas Henry Huxley, was that the advance of human culture moderates the struggle for existence, and should, in time, eliminate it altogether. 'It strikes me,' Huxley wrote, 'that men who are accustomed to contemplate the active or passive extirpation of the weak, the unfortunate, and the superfluous; who justify that conduct on the ground that it has the sanction of the cosmic process, and is the only way of ensuring the progress of the human race ... whose whole lives, therefore, are an education in the noble art of suppressing natural affection and sympathy, are not likely to have any large stock of those commodities left.'[20] There was little natural affection or sympathy in Haeckel when he turned his science to the service of ideology. As Gasman has said, the form that Social Darwinism took in Germany, heavily influenced as it was by Haeckel, was a kind of pantheistic religion of nature worship and nature mysticism combined with racism. As a social organizer, Haeckel is known as the founder in 1906 of the German Monist League, which (until his death) propagated, among other things, the need for racial purity, racial imperialism (*Lebensraum* for the Nordic peoples), and the practice of eugenics.[21]

Haeckel's Monism became familiar to an international reading public with the appearance in 1899 of a short work called (in its English version) *The Riddle of the Universe*. This preposterous, scientifically inaccurate, yet for the uninstructed beguiling[22] tract became one of the biggest 'popular science' best-sellers of all time. It was disliked even by some of Haeckel's closest friends and colleagues. Yet in the first year after publication it sold more than 100,000 copies; it went through ten editions in the next twenty years; and it was eventually translated into more than a dozen languages. It was published in England by the Rationalist Press Association, an organization of agnostic free thinkers, in part because it attacked Christian beliefs. By 1933, half a million copies had been sold in

Germany alone. Thus, Haeckel's dangerous and virulent perversion of Darwin's ideas was widely diffused under the cloak of science and with all the weight of Haeckel's prestige and authority. He represents, in this all-important aspect of his life and work at least, an especially misguided and repellent brand of scientism.

His English disciple Karl Pearson was, as a Social Darwinist, an altogether slighter figure. He was by training a mathematician, but he was also by inclination a philosopher of science who had dabbled in the philosophy of history. Though he is justly famous as one of the founders of the application of statistical methods to evolutionary biology, he appears to have been drawn to this study through an already formed historicistic and Social Darwinist outlook more than by anything else; and his espousal of the eugenics movement, to which he made a major contribution (see next subsection), was likewise an outgrowth of his Social Darwinism rather than its motivation. In an exchange in 1901 with a reviewer of one of his works, he claimed that it was history that had led him to try 'to understand something of inheritance and of racial struggle'; that he had 'turned to the biological field' because he had 'been driven by the force of facts to see that the keynote to the history of man lies in the struggle for food and in the struggle to reproduce, which are the great factors at the base of all biological reasoning.'[23]

Whereas Haeckel's Social Darwinism rested in an original system of ideas, however misguided, Pearson's appears as an uninspired *mélange* of half-digested notions that were currently in vogue in the 'progressive' social circles in which he happened to move. He believed passionately that science must be brought to the service of increasing the welfare of society (through a kind of state socialism); but 'welfare' was to be understood as continued success in the struggle for survival of nation against nation and race against race. The struggle (the existence of which he did not question) was to remain 'among the surviving fit,' a battle against the increasing national degeneracy he saw going on around him. Though he was not an ardent racist, it would be sad, he thought, if 'a false view of human solidarity' should deter 'a capable and stalwart race of white men' from replacing 'a dark-skinned tribe.' Though not strongly anti-Semitic, he believed the Jews were likely always to be 'a parasitic race.' And though not anti-feminist, he urged that a woman's 'rights' must wait upon a sober analysis of her contribution to national fitness. Hence, in *The Woman's* (sic) *Question* (1885) he gave grave consideration to the question of whether the higher education of women might lead to 'a physical degradation of the race, owing to prolonged study having ill-effects on a woman's child-bearing efficiency.'

Darwin's ideas are alive today while Haeckel's are scarcely remembered. The most important elements of Darwinian evolutionary theory are accepted by almost all biologists, and there are some even who willingly accept the label 'neo-Darwinist.' Are there any neo–Social Darwinists? Until comparatively recently one would have been disposed to declare Social Darwinism dead, its doctrines discredited. Writing more than forty years ago in his pioneering work on *Social Darwinism in American Social Thought*, Hofstadter (1944, 1955) did precisely that. Though not entirely ruling out the possibility of a resurgence 'so long as there is a strong element of predacity in society,' he thought that further advances in evolutionary biology would be unlikely to 'affect [future] thought' (1955, concluding chapter). We can no longer make that claim. The 'new science' of sociobiology, which we shall examine in a later section of this chapter, though very different in specifics from the Social Darwinism of the nineteenth century (as one would expect, given the advances in biology since that time), shares many of its basic assumptions; and, politically, it is increasingly drawn upon by movements of the radical right (for example, in France) to justify extremist social policies.

The Modifiability of Human Nature: Eugenics

We shall lose our time,
And all be turned to barnacles, or to apes
With foreheads villanous low.
The Tempest, IV, i, 250

Social Darwinism was a social philosophy, eugenics primarily a social movement that came to prominence in an era of rising middle-class fears for the future of industrial capitalism. Its founder, the Englishman Francis Galton (1822–1911), aspired to make it a science.[24] Eugenics, he wrote in 1883, 'is the science which deals with all the influences that improve the inborn qualities of a race; also with those that develop them to the utmost advantage.' It was an aspiration that led him to leave part of his estate to establish a Chair of Eugenics at University College in the University of London.[25] He was a humane man and cannot be held responsible for the enormities that were later to be advocated and sometimes practised in the name of eugenics; but his aspirations reflected an age-old Utopian dream, for there can be no science of human improvement as such.

Galton was an innovator, a man of restless intelligence. He was the inventor of fingerprinting as a means of positive identification; he was the first to apply statistical methods to the study of heredity in populations; he was the first to put forward the idea of measuring mental ability by 'intelligence tests'; he was an indefatigable measurer, and he was still busy measuring physical characteristics in his old age.[26] He is thus the link between the craniometry of the nineteenth century and the intelligence testing of the twentieth. It was his studies of the family connections of eminent men (which he began in the 1860s) that led him to conclude that not only intelligence, but also moral qualities such as 'character,' were inherited.[27] There followed a series of successful books: *Hereditary Genius* (1869), *English Men of Science* (1874; based on a survey of fellows of the Royal Society), *Inquiries into the Human Faculty* (1883; in which he coined the term 'eugenics'), and *Natural Inheritance* (1889). From this work there emerged the proposal that humans could, and should, be bred selectively 'to improve the inborn qualities of the race.'

This was *positive eugenics*. But there was another side to the story. In the last two decades of the nineteenth century concern was increasingly being expressed about the problem of the fertility of the 'submerged urban poor,' 'the criminal classes,' the mentally defective, and low-status immigrants. Herbert Spencer had argued that natural selection, left to itself, would bring about the gradual elimination of the 'unfit' and thus lead to the continuous improvement of humankind; but by the 1880s this (in any case intolerable) view was becoming less and less plausible. On the contrary, it seemed that, left to itself, natural selection would simply compound social degeneration through the intermarriage of the 'fitter' with the 'less fit' and the swamping of the best by the more rapidly breeding worst. This perception gave a powerful impetus to *negative eugenics*: the protection of 'the inborn qualities of the race' by policies aimed at discouraging the unfit from breeding. This almost certainly meant, in practice, the adoption of coercive measures. Galton (and there were others like him) was reluctant to accept this implication. Apparently accepting the Spencerian view of natural selection, he wrote: 'What Nature does blindly, slowly and ruthlessly, man may do providently, quickly, and kindly.' Though he was prepared, for example, to see habitual criminals permanently segregated 'under merciful surveillance,' he in general preferred education and persuasion. Enthusiasm to improve our race, he wrote in 1901, 'might express itself by granting diplomas to a select class X of young men and women, by encouraging their intermarriages and by promoting the early marriage of girls of that class. The means that are available consist in dowries ... help in emergencies, healthy homes, pressure of public opinion, honours, and the introduction of religious motives, which are very effective as in causing Hindoo girls and most Jewesses to marry young.'[28]

The sociological flow in the eugenic idea as propounded by Galton (there is also a biological flaw in eugenics that I shall mention later) lies in the very title of Galton's essay from which this quotation is taken: 'The Possible Improvement of the Human Breed *Under the Existing Conditions of Law and Sentiment*' (emphasis mine), for it is precisely the conditions of law and sentiment that are likely to prevail in a free society that prevent the aims of eugenics from being realized, except in the very short run – as indeed eugenicists in Britain and America very quickly became aware. The world did not have to wait for a Darwin in order to propose the deliberate manipulation of human breeding. As Medawar wrote, the empirical arts of the stock-breeder are as old as civilization: 'Given a tyrant or a dynasty of tyrants, a scheme of selective inbreeding could have been enforced upon human beings at any time within the last five thousand years' (1972: 72).

The eugenics movement did not take off until the end of the nineteenth century. In Britain[29] its political influence was relatively short-lived and was at its peak between 1901 and the outbreak of the First World War in 1914. It was given institutional expression in the Eugenics Education Society (later, 'Education' was dropped from the title) founded in 1907; and when Pearson was appointed Galton Professor of Eugenics in the University of London in 1911 he established a Eugenics Laboratory there (at University College). The president of the society for nearly twenty years was one of Darwin's sons, Major Leonard Darwin, an army engineer. The movement attracted a coterie of (by the standards of the day) progressively minded middle-class people of diverse interests and connections. Among those associated with it were Havelock Ellis, an early writer on 'sexology,' who contributed an article to the *Eugenics Review* in 1909 entitled 'The Sterilization of the Unfit,' and, in 1911, published a pamphlet called 'The Problem of Race Regeneration' in which he proposed vasectomies for the 'unfit;'[30] Dr Marie Stopes, the leading pioneer of birth control; George Bernard Shaw (at least for a time), who was a Lamarckian and characterized natural selection as a blasphemy; and Ronald Fisher, who eventually succeeded Pearson in the Galton Chair of Eugenics and made fundamental contributions to Mendelian genetics, but as an undergraduate at Cambridge founded the University Undergraduates Eugenics Society (1911), the members of which were to be 'agents of a new phase of evolution' dedicated to spreading 'the doctrine of a new natural ability of worth and blood' (Norton, 1978b). There were also connections with the Fabians in the persons of Beatrice and Sidney Webb and H.G. Wells. An analysis by Mackenzie (1976) of the membership of the council of the Eugenics Education Society in 1914 shows that it included, inter alia, the secretary of the National Association for the Welfare of the Feeble-Minded; the president of the Royal Astronomical Society; the secretary to the Royal Society; Charles Spearman, one of the founders of the 'science' of mental testing; several other scientists, lawyers, and physicians; and W.R. Inge, the Dean of St Paul's Cathedral, popularly known in Britain throughout his well-publicized adult life as 'the gloomy Dean.' In its social and intellectual composition the society was a typically English 'fringe' group of well-intentioned meddlers.

Eugenic ideas were common coin in Britain in the 1920s, but the Eugenics Society itself became less active politically and concentrated more on research and 'education' (that is, propaganda). British research in human genetics was markedly eugenic in orientation in this period, and the then editor of *Nature*, Sir Richard Gregory, gave ample space in its pages to the discussion of eugenic issues. By the early 1930s, however, biologists of a more radical, left-wing temper like J.B.S. Haldane (a long-time member of the British Communist Party), Lancelot Hogben, and Julian Huxley mounted a scientific counter-attack. The cause of eugenics was not helped, moreover, by the defection of George Pitt-Rivers, who had been the secretary of the International Federation of Eugenics Societies, to the British Union of Fascists. Though not moribund (the Eugenics Society still exists, with an office in London), the movement is a spent force except in extreme right-wing circles, its ideas kept alive by fascist organizations such as the National Front (the successor to the British Union of Fascists). It is worth noting that when in 1974 a prominent British politician[31] announced, in an unguarded moment, that the effects of recent immigration from the former colonial empire constituted a threat to 'the balance of our population, our human stock,' it was generally supposed that he had irreparably damaged his political career.

The eugenics movement in the United States shared with its British counterpart an obsession with the protection of the inborn qualities of the human stock, in this case the original (that is, Anglo-Saxon) stock, or 'American type,' from the threat posed by the massing of 'degenerates' and the 'unfit' in the great urban slums. There was, in both countries, the same fit between eugenic ideas and the findings of the rapidly growing mental testing movement. There was the same reliance, for the achievement of many of its aims, on state action. But there were also important differences: greater concern (not surprisingly) in America with the immigrant problem,[32] and the ever-present question of the 'negro race.' Less obviously, perhaps, there was the influence of rural America. It is not without significance that one of the earliest nationwide centres of eugenics developed within the American Breeders' Association, which was founded in 1903 to promote scientific agriculture and the application of the fruits of modern genetic research. In 1906 the association established a eugenics committee and by 1913, when its name was changed to the American Genetic Association, eugenic interests had come to play a prominent part in the organization.[33] There were close links between some strands of the eugenics movement in America and the promotion of a revitalized rural life. One of the major promoters of the American Breeders' Association and a eugenics supporter, the then assistant secretary of agriculture Willet Hays, saw the farm as 'the home of the race,'[34] and the improvement of rural America as the key to racial regeneration. The notion that rural living could counteract the dysgenic effect of urban squalor can be seen in the methods proposed by the Movement for Racial Betterment. To the more general and familiar eugenic aims of 'total abstinence,' sterilization of defectives, and 'eugenic marriages,' there were added 'simple and natural habits of life,' fresh-air schools, and 'back to the farm' policies.

The chief difference between the movements in the two countries lies, however, in the much greater political influence that was exercised by eugenicists in the United States. In 1907 the State of Indiana passed a sterilization law aimed at defectives and criminals, and this example was quickly followed in fifteen other states, and eventually in twenty-one. Anti-miscegenation measures were on the statute books of several states by 1930. At the national level, probably the greatest lobbying triumph of the eugenicists was the passage of the Johnson Act by the United States Congress in 1924. This act effectively blocked further immigration from Asian countries, and severely limited immigration from southern and eastern Europe.

It is ironic that one of the examples of supposed hereditary degeneration most often used by American eugenicists to support their policies was drawn from rural life. This is the case of a family, the descendants of five sisters and a single father, who lived in the Finger Lakes region in upstate New York in the second half of the nineteenth century. The case histories of this family (disguised as 'the Jukes') were compiled by Richard Dugdale, a penal reformer, and were published first as a report to the New York State Prison Association in 1874 and later as a book for the general public. What is important, however, is not so much the rural origins of the family as the subsequent history of Dugdale's report. This history illustrates, once again, the extent to which evidence can be distorted by preconceptions and by unconscious (and sometimes – as is evident here – not so unconscious) biases. Dugdale came to be represented as a hereditarian, whereas he was, in fact, persuaded that the misfortunes of the Jukes stemmed not from their heredity but from their environment. For example, a scientist president of the University of Indiana and later the first president of Stanford University, David Jordon, wrote in his book *Footnotes to Evolution*, published in 1898, that 'Dugdale had shown' that crime and poverty reappear generation after generation, concluding that 'every family of Jukes' that enters the United States 'carries with it the germs of pauperism and crime.' C.B. Davenport, a prominent eugenicist member of the American Breeders' Association, and the founder in 1910, with money provided by the Harriman family, of a Eugenics Record Office at Cold Spring Harbour, was another Dugdale misrepresenter. His organization claimed to be dedicated to the pursuit of 'precise data collection'; but this did not deter him from asserting that the Jukes were families of defectives and criminals, paupers, and sexual deviants, all tracing back to a single ancestor. As a report in *The Scientific American*[35] (from which these facts are taken) puts it: 'By the early years of the 20th century the hereditarian interpretation of the Jukes studies had become established as an argument supporting the view that poverty, crime, and a variety of other social problems are the result of inborn tendencies.' This interpretation gave support to eugenicist policies. Yet, 'Dugdale's proposed remedies had been quite different: he had recommended improved health care, remedial education for the underprivileged, penal reform, and assistance to poor families.'[36]

In Germany, the membership of the Monist League included many hard-line eugenicists. Haeckel was their inspiration. As Gasman has shown, some of them even believed that entire social classes could possess unique hereditary characteristics that made some 'weak' and others 'strong.' Dr Wilhelm Schallmeyer, an important figure in the Monist League and one of Germany's foremost promoters of eugenic ideas, argued that 'biology' held the key to the rise and fall of civilizations (Gasman 1971: 91). It was therefore the duty of the state to ensure the reproduction and survival of only the fittest individuals. Childless couples (of the right type) should be forced to remarry in order to carry out their duty to the nation; it should be a criminal offence to remain unmarried; persons of the upper classes ('fit' by definition) were guilty of a dereliction of duty if they married late. Heinrich Ziegler, a co-founder with Haeckel of the Monist League, was appalled by the number of 'defectives' in German society, assuring his readers that the feeble-minded committed most of the crimes, and were guilty of most of the drunkenness, and that most murderers were feeble-minded or epileptic. The vice-president of the league, one Johannes Unold, wrote hysterically of the degeneration brought about by indulgence in alcohol, illicit sexual relations, and the reading of pornography, all of which would lead to a weakening of 'the life force.' These Monists were blandly prepared to advance policies of the most appalling cruelty on the slenderest of evidence. Haeckel wrote that infanticide should not be considered a crime when the child was 'abnormal'; rather it was 'a practice of advantage both to the infants destroyed and to the community.' Individuals with chronic diseases, Schallmeyer wrote, should 'probably' be destroyed, and certainly should not be permitted to reproduce. Haeckel proposed, in his sequel to *The Riddle of the Universe*, inappropriately entitled *The Wonders of Life*, that there should be established a commission to decide whether the ill and deformed should live or die. If the latter, the unfortunate souls should be dispatched by doses 'of some painless and rapid poison.'

Though their main emphasis was on negative eugenics – on the elimination of 'degenerates' and 'undesirables' – consideration was also given to positive eugenics. This meant, in practice, encouraging the 'best' classes to breed more prolifically. And this, as Gasman has shown (1971: 99), raised the question of the emancipation of women. To the more liberal-minded rank-and-file members of the Monist League (of whom there were many) the emancipation of women meant, within the constraints of the social opinion of their time and place, precisely what it said. The leadership of the league, far to the right politically, saw it quite differently. For them, 'emancipation' meant freeing women to fulfil 'their own special natures.' It is scarcely necessary to specify what these were. The way to fulfilment consisted in doing their duty to the state: having many children, being 'good' mothers, and maintaining a sound bourgeois family life. Emancipation meant, not minimizing the differences between the social roles of men and women, but rather removing hindrances to the accomplishment by women of their 'proper' biological functions. To go further than this would lead to 'racial death.'

In most of the foregoing discussion in this section questions of morality have never been far from the surface. There is, however, another dimension to eugenics that must now be noted. There is, as I said earlier, a fatal biological flaw in the eugenic idea that completely destroys its hopes and makes its ultimate aims impossible to achieve. This flaw could not have been known to the eugenicists of the day for it is only comparatively recently that advances in biological understanding have made it plain. It has been clearly explained by Medawar (1972: 72–6, 91–5). What follows is a very brief summary of his argument.

Until recently, stock-breeding (on which analogy positive eugenics rests) was expected, and believed, to fulfil two quite different aims in the same process. The first was to produce some desired result in the offspring of the breeding individuals – for example, more splendid racehorses, or cows that produced higher yields of milk; the second was to be able to use the produced offspring as parents of the next generation. In order to meet this second aim (the reproductive aim) it was necessary for the second generation of parents to be so constituted genetically that they would be true-breeding (hence 'thoroughbreds') in respect of *all* the characters for which the selection had been originally undertaken. It is as though, as Medawar puts it, Rolls Royces, in addition to being the end-product of manufacture, had to be so designed as to give rise to Rolls Royce progeny. It is now known why these two aims of the stock-breeder cannot be combined in the same process. Individuals do not breed true; their offspring are always unlike them genetically in many ways. It is populations as a whole that breed true. 'The end product of evolution [within a species], in so far as it can be said to have one, is itself a population, not a representative genetic type to which every individual will represent a more or less faithful approximation. The individual members of the population differ from one another [genetically] but the population itself has a stable genetic structure i.e. a stable pattern of genetic inequality' (Medawar: 74). What stock-breeders do in practice is to follow elaborate programs of cross-breeding in order to continue to produce the kind of results they wish to achieve. This means that their practices can no longer be advanced as an analogue of policies of positive human eugenics unless those policies were to utilize means more draconian even than those that must inevitably apply to policies of negative eugenics.[37] Medawar stresses the moral objection: that no policy of positive eugenics requiring such means for its execution would be tolerable in a society respecting the rights of individuals. I would argue further that the necessary means would be impossible to apply in even the most totalitarian of complex societies, since the degree of interference and control required to implement them would be unattainable.[38]

Finally, we should not forget that the hereditarian assumption on which the entire edifice of eugenics rested – the assumption that intelligence and even moral qualities are, in themselves, fixed and unalterable characteristics transmitted by inheritance – was, and remains, an assumption that is based on the slenderest of scientific evidence. Unfortunately, it dies hard; and it is to this issue that we now turn, in the first of our two case studies.

THE IQ CONTROVERSY: THE CASE OF ARTHUR JENSEN

If you cannot measure, measure anyhow.
Frank Knight, economist

What came to be labelled 'the IQ controversy'[39] began with the publication in 1969 of an article by the educational psychologist Arthur Jensen in the *Harvard Educational Review* (Jensen 1969). This 124-page essay was essentially nothing more than a 'review of the literature'; but the item that attracted most attention was Jensen's assertion that (American) blacks are intellectually inferior to (American) whites because, on average, blacks score fifteen points lower than whites on standardized tests of intelligence, and statistical analysis had shown that 'to a very large extent' (the extent that has been claimed varies between 60 and 80 per cent) this inferiority was inherited.[40] The supposed implications for policy were that attempts to raise the intellectual capabilities of blacks through special education were futile and a waste of public money, and that, given their higher fertility rate, blacks constituted a threat via their 'inferior genes' to the quality of the future population of the United States. What Jensen set out to do in the article was to challenge the basis of the policy that had guided the compensatory education projects that were then in place in the black urban ghettos.

There was nothing particularly new or original about the 'facts' in Jensen's article; what was interesting was its timing. Earlier attempts to revive the hereditary basis of intelligence as a public issue and to demonstrate its implications for human inequality, supposedly on the basis of the findings of IQ psychology (such as the book by Audrey Shuey, *The*

Testing of Negro Intelligence, published in 1958) had come to nothing. The moment Jensen chose to publish coincided with a particularly high point of social unrest and racial tension in the United States. Robert Kennedy and Martin Luther King had been assassinated the previous year; the Black Panthers, an organization of blacks dedicated to violence and armed self-protection, had recently been formed; radical student opposition to American involvement in Vietnam (interpreted as racist imperialism) was at its height and had spread to Europe; in the year of publication (1969) there took place the notorious Chicago conspiracy trial. It is also noteworthy that Jensen had seen fit to release a copy of his article to a journalist, a staff writer for the *U.S. News and World Report, before* it appeared in the *Harvard Educational Review*. This suggests not only that Jensen was aware of the broad public impact that it might make, but also that he was not wholly disinterested in the results of his action.

For more than a quarter of a century, from the mid-1930s, 'scientific racism' lay dormant as a public issue. Indeed, when the full extent of the atrocities perpetrated by the Nazis in the name of race became generally known at the end of the Second World War, many people assumed that it had been banished for good. This assumption was not unjustified, since arguments about the racial purity and moral superiority of whites could scarcely be expected to survive the opening of the Nazi death camps by the advancing Allied Armies in 1945. But it was not to be.

The cries of outrage and the charges and countercharges that followed the publication of Jensen's article were predictable. Jensen was supported in the United States by, among others, Richard Herrnstein, a Harvard psychologist (who extended the debate to class differences as well as black / white differences), and William Shockley, an engineer and one of the inventors of the transistor, for which he shared a Nobel Prize for physics in 1956. Jensen's leading advocate in Britain was H.J. Eysenck, also a psychologist, who published a book called *Race, Intelligence and Education* in 1971. Ranged against them were R.C. Lewontin, a biologist, David McLelland, a Harvard psychologist, Leon Kamin, another psychologist at Princeton, Noam Chomsky, and other lesser lights. Generally speaking, the main battle was joined between a group of educational psychologists on one side and a mixed group of biologists and psychologists on the other; but before the fighting waned scientists from many disciplines became involved, along with social scientists, philosophers, and publicists and journalists (for the fight went quickly beyond the confines of academia). At one point the controversy became quite vicious, and (among other well-publicized occasions) student activists prevented

Eysenck, by force, from speaking at the London School of Economics, at the University of Leeds, and at the University of Sydney, Australia (where demonstrators tried to break up the meeting with smoke-bombs), and Shockley from speaking at Berkeley, Dartmouth College, and other universities; and the proposed award of an honorary degree to Shockley for his scientific work was withdrawn at the last moment by the Senate of the University of Leeds as a result of strong pressure from faculty and students. Jensen was the object of demonstrations on the campus of his own university and at other universities in the United States at which he was an invited lecturer; and when the International Congress of Genetics met at the University of California at Berkeley in 1973, arriving delegates were greeted by a sign reading: 'Welcome to Berkeley, Home of A. Jensen, Professor of Racism.'[41]

Not the least interesting of the many human aspects of this case is the personal interrelationships that existed among the protagonists on the 'hereditarian' side. Jensen had studied under Eysenck as a graduate student; Eysenck was a student of Sir Cyril Burt[42] and was considered by Burt to be his brightest pupil (he completed his PhD dissertation under Burt's supervision in 1940); Burt had been a pupil of Karl Pearson, who, in turn, was strongly influenced by Galton. It therefore makes sense to talk, as does Steven Rose (1976: 130), of an invisible college of 'mental testers,' linked not only cognitively but socially as well, extending over a period of more than a hundred years.

The emotional and highly charged political atmosphere surrounding the case was unquestionably heightened by the presence of Shockley, who had no professional qualifications whatever for speaking to the issue, but whose fame as a Nobel laureate may have been expected to add some weight to his opinions among the general public. In fact, while Jensen, however misguided, appeared to be a dedicated scholar sincerely convinced of his views, Shockley's extravagant pronouncements seemed to condemn him as a racist of the worst kind; for example, his statements that 'intelligence is colour-coded' and that the higher IQ scores of some blacks relative to other blacks shows that they 'have more Caucasian blood.' He, of course, turned the racist epithet around and accused his critics of racism on the ground that they wished to discriminate against whites. He once claimed, in an extraordinary Messianic outburst, that he was using his talents 'in keeping with the objectives of Nobel's will: of conferring greater benefits on mankind'; and, in an ingenious exhibition of perverse logic when his request for funds to investigate the claim that intelligence is 'eighty percent inherited' was turned down on the grounds that it was improperly motivated, charged the National Academy of Sciences with 'Lysenkoism' and with 'the irresponsibility that existed in Nazi Germany when intellectuals were unwilling to face up to the horrors that might be happening to the Jews.'[43] (He got some money anyway, but from an organization called the Pioneer Fund for Racial Betterment.) He claimed that his view that the Negro was 'inherently less intelligent' than the Caucasian came from his 'concern with the welfare of humanity,' and that pointing such things out made him 'the intellectual in America most likely to reduce Negro agony in the next generation.' His proposals for 'dysgenics' were no different in principle from those put forward by old-time eugenicists. They included sterilization, but with cash inducements, the amount of which would vary in inverse proportion to the IQ score of the person accepting sterilization.

Jensen and his supporters complained incessantly that they had been denied their rights as scientists to publish and speak about their work and to be free 'to pursue the truth wherever it may lead'; and in 1972 they published a kind of manifesto, a 'Resolution on Scientific Freedom,' to this effect.[44] The press tended to be sympathetic. For example, when Eysenck had his eyeglasses broken in a scuffle at the London School of Economics, the *Guardian* newspaper (formerly *Manchester Guardian*) commented editorially that all Eysenck wanted was 'to be left in peace to advance his research in genetics.'[45] While mob rule is to be deplored, this really will not do. Nor, in this context, will an argument from the other side, that scientists should refrain from publishing findings that are politically and socially sensitive.[46] On the basis of that argument the Jensenites had right on their side, but both are red herrings. Whatever may have been Jensen's original motive for pre-releasing his 1969 article to the press, neither he subsequently, nor any of his supporters at any time, showed reluctance in 'going public.' Indeed, they deliberately sought to do so. Richard Herrnstein, for example, made his first contribution to the controversy, not in an academic journal, but in the *Atlantic Monthly* (September 1971).[47] What the Jensenites did was to take a body of scientific work out of the scientific arena in which they were subject to the legitimate criticism of their peers, and appeal over the heads of their opponents to the public for support. But public opinion cannot settle a scientific question. This is not to say that other scientists have never been guilty of similar practices. Nor should one condone the attempts that were made to silence Jensen and his supporters by physical violence, intimidation, and shouting them down. Not only were these incidents inexcusable in themselves, they gave the Jensenites the look of martyrs and an opportunity to present themselves as latter-day Galileos locked in combat with the forces of unreason.[48]

There appear to be five major components of 'Jensenism.'[49] First, that intelligence is a specific, objective, 'faculty of the mind' – a single 'thing in itself.' Secondly, that it can be measured, if not directly (like the skull or the brain inside the skull), then indirectly.[50] Thirdly, that it can be measured by objective tests (intelligence tests) that produce a single-number valuation, the intelligence quotient or IQ. Fourthly, that multi-factoral statistical analysis of the results of thousands of these tests leaves no doubt that intelligence is in large measure inherited and relatively unaffected by environmental factors.[51] Fifthly, in specific reference to race (leaving aside the matter of social class, and groups within a class or race such as men and women or the Irish in North America,[52] to which the same has been held to apply), since the analysis shows that some races are more intelligent on average than others, racial inequality is 'natural' not sociological. All these components have been contested at great length, and often with devastating effect, since the Jensen case first 'broke.'[53] I shall mention only a few of the more important criticisms that have been made.

The notion that intelligence can be treated as if it were an objective 'thing' has been vigorously disputed by Lewontin (1981). Intelligence, he argues, like acquisitiveness or moral rectitude or saintliness (and, we might add, since the craniological forerunners of IQ psychology made so much of it, like anatomical beauty), is not a 'thing' but a mental construct and as such is historically and culturally contingent.[54] To treat it as a thing is to reify it. It might be argued that Lewontin is claiming too much here, but it is worth noting that the French psychologist Alfred Binet (1857–1911), who was the first to use intelligence tests (to identify retarded children needing special educational treatment), defined intelligence very cautiously as the totality of mental processes involved in adapting to the environment – which would at least argue for cultural and historical contingency, since environments differ, and differ over time. Unfortunately Binet's caution has not been respected by his successors.

Secondly, what is it that intelligence tests measure? Jensenites argue that the multifactoral analysis of IQ scores produces a single measure, an entity, which they call general intelligence, or 'g.' As Gould (1981a) puts it, reification of IQ as a biological entity has depended on the conviction that 'g' measures a single, scalable, fundamental 'thing' residing in the human brain. 'I will at least say this for Arthur Jensen: he recognises that his hereditarian theory of IQ depends upon the validity [of the concept] of g' (p. 256). Gould then devotes nearly one hundred pages to a technical dissection of the invalidity of 'g,' and he is one of the first people outside the profession of psychometrics to have done so. 'G' is a measure abstracted from correlation coefficients (the theoretical rationale for 'g' is that the results of many *different* types of intelligence tests positively correlate) but, says Gould (1980b: 42), 'the oldest truism in statistics' is that correlation does not imply cause.[55] 'Even if dominant g were an ineluctable abstraction from the correlation matrix [which Gould denies] it still wouldn't tell us *why* mental tests tend to be positively correlated. The reason might be largely innate as Jensen assumes and requires: people who do well on one test usually do well on others because they have more inborn intelligence. Or it might be largely or totally environmental ... because they had a good education, enough to eat, intellectual stimulation at home, and so forth.'

Since Jensen has admitted (in an exchange with Lewontin) that IQ (which he equates with abstract reasoning ability) 'is ... a selection of just one portion of the total spectrum of human mental abilities' and that 'other mental abilities have not yet been adequately measured' (Block and Dworkin 1977: 97) it would seem that he should also admit that intelligence is not something to which a single-number valuation can be attached. Single-number valuation of complex variables, Medawar wrote (1977a), was once tried in demography, and measures of a nation's 'reproductive prowess' were developed; but today no serious demographer believes that a single-number valuation of reproductive vitality is possible. The reason is that there are too many variables and not all of them are scalar in character. They include (among scores of others) the proportions of married and unmarried mothers in the population, prevailing fashions about the desirable size of a family and about the age of marriage, fluctuations in the state of employment, taxation policy, and the availability and social acceptability of methods of birth control. Similarly in soil physics, where it has proved impossible 'to epitomise in a single figure the field behaviour of a soil'; and in macroeconomics, where some economists have tried to use growth in GNP (or GDP) as a measure of national welfare. Any such use, Medawar wrote, is 'totally inadmissible' since no single figure can evaluate such things as a nation's confidence in itself, its concern for the welfare of its citizens, the safety of its streets, and its political stability. He adds: 'IQ psychologists would nevertheless like us to believe that such considerations as these do not apply to them.'

Medawar goes on to point to 'a still graver illusion,' the illusion that intelligence tests can be devised that are not culturally biased. This is familiar ground and, of course, Jensenites deny bias, usually by claiming that cultural bias can be 'factored out.' Jensen has argued, for example, that the claim of culture bias 'as an explanation of the White-Negro difference in the United States runs into numerous difficulties. For one thing, many of the tests that show the greatest White-Negro difference show much smaller differences for other minority groups which are also regarded as disadvantaged or culturally different.'[56] As Gould (1980b) has shown, Jensen's book *Bias in Mental Testing* (which has been widely assumed to provide the 'proof' that mental testing is unbiased) is largely devoted to demonstrating that *statistical* bias does not affect the results of mental tests. This, though correct, confuses the issue. What most critics mean by bias in mental testing is not the purely technical matter of statistical bias but bias in the vernacular sense, which is something like 'unfairness of treatment.' Jensen does admit in his book that the technical term 'bias' should be distinguished from the concept of fairness (that is, he admits that when he uses 'bias' he means statistical bias), but he buries 'these brave words' in two pages of an eight-hundred-page book and makes no further reference to them (Gould 1980b: 39), while continuing to use bias throughout the book in the vernacular sense. Gould stops just short of asserting that this confusion is deliberate, but remarks that 'non-statistical' readers of the book will be confused nonetheless. In fact, the evidence is strong that cultural bias in mental testing does exist; indeed it is virtually certain that an entirely culture-neutral test (that is, one that can be applied without bias in *any* culture) is impossible to devise.[57]

Even were it to be agreed, however, that intelligence is a specific 'thing' or faculty of the mind, and that it can be measured by culturally neutral, objective intelligence tests, and assigned a single-number valuation, could the sophisticated statistical analyses of the Jensenites tell us anything about the hereditable inequality between races, classes, and other social groupings? The answer from genetics would appear to be 'No.' As Layzer (1977: 201) points out, geneticists have been clear 'for well over half a century that it is meaningless to try to separate genetic and environmental contributions to measured differences between different strains bred under different environmental conditions.' In this context, 'strains' would mean ethnic and socio-economic groups, and there are clearly *systemic* differences in developmental conditions – cultural, linguistic, and environmental – between these groups that are known to influence performance in IQ tests. 'Since we have no way of correcting test scores for these differences, the only objectively correct statement that can be made on this subject is [that] reported differences in IQ tell us nothing whatever about any average genetic differences that may exist.' In fact (though this is not a point made by Layzer), studies by geneticists of genetic differences, such as blood types within and between 'races,' show that at least 60 to 70 per cent of the overall genetic variation in humankind occurs *within* races.[58] Thus, in effect, 'race' becomes a statistical concept – a quality of a population with such differences as blood type – but blood type is a genuine genetic entity, unlike IQ.

The Jensen affair posed a serious threat to the public image of 'mental testing,' much of which is legitimate and soundly based. The dogmatic public claims of Jensen and his followers led to a more careful, wide-ranging, and detailed re-examination of its assumptions and methodology than had ever been done before. In one important direction this re-examination led to the exposure of the manipulations and methodological errors of Sir Cyril Burt, whose studies of the intelligence of identical twins separated at birth had hitherto constituted one of the foundations of the hypothesis that intelligence is largely inherited and little affected by social and environmental factors. Leon Kamin of Princeton University published his exposé of Burt's work on twins in *The Science and Politics of I.Q.* (1974) and P.D. Dorfman did the same for Burt's equally famous fifty-year study of the intelligence of 40,000 pairs of fathers and sons in a London borough in *Science* (vol. 201, no. 4362, 29 September 1978, pp. 1177–86). To Jensen's credit, he also worked on Burt's twins data at about the same time as Kamin and was puzzled by what he discovered. He concluded in an article in *Behaviour Genetics* in 1974 that Burt's correlations were 'useless' adding the cautious but revealing remark, 'It is almost as if Burt regarded the actual data as merely an incidental backdrop for the illustration of the theoretical issues in quantitative genetics, which to him seemed always to hold the centre of the stage.'[59] Which is a strange quality to ascribe to someone working in a tradition that holds that 'facts' are more important than theories. In fact, Jensen is saying elliptically what is almost certainly true: that Burt spent his life attempting to find the evidence that would support what he 'knew' already, namely that intelligence is innate. From that conclusion, nothing could shake him. This is a much more important conclusion to draw than the still-disputed one (which can probably never be settled conclusively) that Burt was guilty of deliberate fraud.[60] Barry Barnes was quoted earlier (note 48) as saying that the opponents of

Jensenism were seeking 'to defend an occupational privilege.' This could be said far more aptly about Burt. Hearnshaw (1979) has written that controversy with his critics became one of Burt's major activities and motivated much of his work during his declining years; that it was almost as if he had marked out a certain territory for himself within the boundaries of which he was determined to retain the mastery, lay down the law, and drive off all rivals. In fact, what motivates the opponents of Jensenism is not the defence of occupational privilege (an implausible charge anyway since they are drawn from so many different disciplines), but the weight of the evidence.

Much of this evidence comes from modern genetics, especially population genetics, but some comes from within the discipline of psychology itself, notably from Piaget's work and that of other learning theorists, which suggests strongly that intelligence is not a simple single-factor attribute but a complex of information-processing skills learned through experience and trial and error. There are also numerous studies in psychology that have produced results entirely inconsistent with the hereditarian thesis: for example, studies of the effects of different child-rearing practices and of nutrition on differentials in IQ, studies of illegitimate children that strongly suggest that within-family resemblances in IQ are largely the result of similar environments, and studies of the IQs of adopted children and their biological parents that have shown large differences between the two sets of scores.[61] Layzer (1977: 236) draws attention to the fact that many tests have shown that blacks living in the urban north of the United States score systematically higher on IQ tests than those living in the rural south. For many years hereditarians argued (a typical hereditarian response) that this could be explained by the fact that the emigrants (from south to north) could generally be expected to be more intelligent and resourceful than those who stayed at home. Further studies showed, however, that the IQs of migrant children increased systematically and substantially with length of residence in the north. The response of the Jensenists to such counter-evidence is either to dismiss it, or explain it away.

Jensen has continually admonished his critics to learn more about genetics and to 'think genetically' (whatever that may mean); that is, he assumes that their problem (apart from their political motivations, which he also stresses) is simple ignorance, and that with a better education they would see the light. But, as Gould has pointed out (1981a: 318), Jensen not only shows a shaky acquaintance with genetics himself, but also displays an astonishing ignorance of the rudiments of evolutionary biology, believing evolution to constitute a unilinear progressive sequence from 'lowest' to 'highest.' For example, he writes in *Bias in Mental Testing* that chickens, dogs, monkeys, and chimpanzees 'are roughly scalable along a "g" dimension ... "g" can be viewed as an interspecies concept with a broad biological base culminating in the primates' (Jensen 1979: 251). This 'caricature of evolution,' to use Gould's expression, demonstrates as clearly as it is possible to do the Jensenist perpetuation of the nineteenth-century obsession with the idea *that characteristics must be ranked.*

It has been argued by the more radical critics of Jensenism – for example by Leon Kamin, Steven Rose, and Richard Lewontin – that IQ psychology is simply ideology dressed as science. Lewontin writes, in specific reference to Jensen: 'Jensen's article [1969] is not an objective empirical scientific paper [cf. Barnes's statement that it is "heavily polluted with policy considerations"!] which stands or falls on the correctness of his calculation of hereditability. It is, rather, a closely reasoned ideological document springing ... from deep-seated professional bias and permeated ... with an elitist and competitive world-view' (1977: 108). Rose has written that, whereas in the United States mental testing was from the start linked to two explicit goals, the control of foreign immigration (it was so used by legislators) and the suppression of black Americans (it was also used to justify state sterilization laws for the 'mentally unfit'), in Britain it was linked to social class, and in particular to the general support of meritocratic ideas that were congenial to the professional middle classes and, perhaps to a lesser extent, the business elites (Rose 1976: 136). For a while, following the traumas of Nazi racism and of 'the end of empire,' there was an ideological truce, but this was to be shattered by events of the sixties and early seventies: in the United States by the resurgence of racial tensions and increasing concern about the cost of poverty and educational programs, and in Britain by growing disillusionment with the education system and the welfare state on the part of the middle classes, and concern among all social classes about the influx of non-white immigrant labour from the former colonial empire. In brief, the re-emergence in a new dress of old ideas about race and class was as much a response to the 'social contradictions' in modern society as Social Darwinism and craniometry had been to 'social contradictions' in the nineteenth century. IQ psychology, like its antecedents, provides an apparent rationale for the existing social order. Whereas social inequality was once attributed to the will of God, it is now portrayed as conforming to a biological imperative.

There are merits in such arguments for ideology, but I do not believe that this is the whole story. For one thing the argument for ideology cuts both ways. The IQ psychologists believe that criticism of *their* ideas is 'socially motivated.' Some contemporary historians and sociologists of science, indeed, hold that all science is ideological, in the sense that it takes place in specific cultural milieux (which is trivially true) and thus (which is not) *necessarily* reflects prevailing social interests. According to one of them, 'Science as an intellectual formation and ideology as world-view can never be separate realities or autonomous "things" merely interacting, but must also be mutually constitutive of each other or interpenetrating to form a seamless web' (Cooter 1980: 239). Scientific truth, according to such accounts, is a cultural artefact, socially negotiated and organized; and the reality that science describes is 'socially constructed.' There is no objective truth. It follows that there are no external criteria for distinguishing 'good' science from 'bad' science; indeed, 'what does the focus on "bad" science accomplish, but the obfuscation of the ideological power of "good" science?' (p. 260). I have already given reasons for rejecting the view that science is nothing but a cultural process (though it is that too). It follows that I do not accept Cooter's arguments as the whole story either. My own belief is that we cannot ignore, in any analysis of the curious and in many respects repellent phenomenon of 'Jensenism,' the presence of a misguided methodology, and an erroneous view of the nature of science that leads inevitably to its perversion in practice. It seems important to ask why, if a careful critic like Kamin thought that Jensenism could be dismissed as purely ideological, he would have devoted so much patient effort to an exhaustive reworking and highly effective demolition of the evidence and reasoning on which Jensenist arguments are based.

Jensenism is the latest (it is unlikely to be the last) of a long line of doctrines that base their claim to scientificity on respect for 'hard data' and measurement. One of Jensen's more memorable pronouncements is that 'the most important thing about intelligence is that we can measure it' and that there is no point in arguing about what it 'really' is. This is a strange statement; not because we *do* know what intelligence is (in fact 'intelligence' may simply be an unnecessary and misleading concept), but because of the statement's outright dismissal of explanatory theory, one of the most important aspects of science. Jensenists are quite happy to define intelligence operationally. Intelligence is what intelligence tests measure, as the psychologist E.G. Boring once said. Jensen accepts this definition. In their operationalism they are following the American physicist Percy Bridgman (who gave operationalism its name). Bridgman argued that it should be possible to reduce all the concepts of a science to accounts of practical measuring operations.[62] For example, 'length' could be defined by the set of operations by which length is actually measured. But operationalism has long since been shown to be invalid. It overlooks among other things the fact that operational definitions must necessarily be circular. For example, an operational definition of 'length' would have to allow for temperature variations, but the operational definition of 'temperature' involves measurements of length (as in a thermometer). Operationalists argue that their definitions are purely conventional; in this case, that it is agreed that when we mental testers say 'intelligence' we *mean* only 'that which intelligence tests measure' and imply nothing more than that. It is a purely operational definition. But if that is so, it is circular.

The Jensenists are heirs to the naive inductivist supposition of their craniometric predecessors, that science proceeds by the processing of facts. But there are no 'facts' in the absence of a theoretical framework within which observations and experimental results *become* facts, a framework within which they are selected and interpreted. Jensenist theories are crude and only vaguely acknowledged – for example, that superior intelligence is shown by 'success in life' and that lack of success in life means inferior intelligence – but they are theories nonetheless. The Jensenists' disdain for theory is evident, however, in their belief that a scientific study of human behaviour should make do with the weakest possible hypotheses. This belief is held by some of their antagonists also, notably by the more radical of the environmentalists. Kamin, for example, asserts that the weakest possible hypothesis is that all observed differences in IQ are attributable to *non-genetic* factors and that the data do not compel us to abandon this hypothesis. But Jensen asserts precisely the same of *genetic* factors! This antiquated doctrine holds, in effect, that investigation must start from a position of disbelief (the theory should make the fewest possible assumptions) and that we then strengthen our position by the successive accumulation of empirical reasons for abandoning our disbelief.

It is, however, a serious mistake in method to advance the weakest hypothesis because this is the least testable, that is to say, the least falsifiable. It can be confirmed, but confirmations do not constitute a genuine test of any theory.[63] Jensen has said that, 'in the absence of any compelling environmental explanation for the white-negro intelligence difference, it would be scientifically remiss not to seriously consider the genetic hypothesis.' Though this statement is no doubt intended to be evasive ('remiss not to seriously consider'), it is quite clear, in the context of Jensenist *practice*, what it means. It means that if we successively eliminate all environmental factors one by one we shall have established the contrary, the genetic hypothesis. But this is invalid. Two contrary hypotheses may *both* be false.[64] In this case they probably *are* both false, since much present evidence suggests that, whatever 'intelligence' may be, it is the result of interaction between genetically inherited propensities and a fluid (or 'plastic') environment. The issue of the relative importance of heredity and environment is a prominent feature also in the recent controversies over sociobiology, to which I now turn.

THE 'NEW SCIENCE' OF SOCIOBIOLOGY

Hyena pisses from fear so does man and so does dog.
Charles Darwin, the 'Man' notebooks

For more than a hundred years it has been impossible to talk about the nature of human nature without taking account of Darwin's shocking revelation – our animal origins. Although the earlier Darwinian belief that we are 'descended from the apes' has been shown to be not strictly correct (our hominid ancestors said goodbye to them some eight to ten million – some estimates give four to seven million – years ago) it is no longer possible for us to deny that we, in common with all other living things, are the products of an immensely long biological evolution. But to what extent are we *determined* by that process? It is this question that lies at the heart of the often rancorous dispute that began in 1975 with the publication of Edward O. Wilson's book *Sociobiology: The New Synthesis* (Wilson 1975).

The nub of the dispute can be stated very simply. Sociobiologists seek to ascribe almost all our behavioural differences from other animals to the same sources from which we derive our behavioural similarities: to our genetic heritage and the working out of the processes of natural selection. The counter-argument is that we are now free of what Medawar called 'the tyranny of our biological inheritance' (1977a) by virtue of the fact that the principal modality of human evolution is no longer genetic but 'psychosocial' (to use Julian Huxley's term) or, as Medawar prefers, 'exosomatic.' Ordinary organic evolution, says Medawar, 'is mediated through a genetic mechanism, but exosomatic evolution is made possible by the transfer of information from one generation to the next through non-genetic channels' (1977b: 53); in other words, through cultural pathways created by us and external to our biological selves. One of the most important of these pathways is human language. The Lamarckian inheritance of acquired characteristics has no longer any place in biology, but it is precisely the way in which cultural evolution takes place.[65] Learned behaviour is transmitted culturally, and a characteristic is therefore 'acquired' by being passed on through successive generations.

Edward Wilson's central thesis is that all the characteristics that we think of as distinctively human, such as altruism, morality, and creativity, and almost all our institutions such as religion, are – directly or indirectly – genetic in origin.[66] He does not deny a place for culture but, like Jensen, he gives it only a small place. The hereditary component, he says, is decisive. It is already necessary, however, to enter a caveat. A statement about the genetic origin of a characteristic does not necessarily tell us much about (let alone explain) its present function. In the case of the social insects, for example, it may do; but when we turn to cultures as highly sophisticated as human cultures this is most unlikely. The counter (and much more plausible) hypothesis to Wilson's is that although we evolved from earlier hominid forms mainly genetically, our subsequent establishment as fully cultural beings has been achieved through non-genetic pathways.

The working out of Wilson's thesis was begun in the last chapter of his book *The Insect Societies* (on which subject he is a world authority), which he published in 1971. In that chapter, 'The Prospect for a Unified Sociobiology,' he argued that the principles that had worked so well in explaining the rigid systems of the social insects could be applied point by point to the more flexible systems of vertebrate animals. This is exactly what he proceeded to do in 1975 in *Sociobiology: The New Synthesis*, a massive work of extraordinary erudition. The last chapter of the book, 'Man: From Sociology to Sociobiology,' not only proposed but actually began to extend the same principles to human societies. This chapter was clearly intended to be noticed by social scientists since it began by arguing that one of the functions of the new discipline of sociobiology was 'to reformulate the foundations of the social sciences,' by which it was evident he meant not only sociology, anthropology, and politics, but philosophy and ethics as well.

It is not surprising that this audacious proposal for a research program, at once thoroughly reductionist in aim and encyclopaedic in scope, should have provoked a hostile reaction. Social scientists and humanists saw it as a 'take-over bid' – correctly, since Wilson had written that sociology and the other social sciences 'as well as the humanities' were the last branches of *biology* waiting to be included in the new synthesis (1975: 4). Wilson seems to have been bemused as well as hurt by the strength of the reponse; but, as one commentator put it, the ingenuous blend of arrogance and condescension with which his program was presented made it almost inevitable. It did not restrain his ambitions however. Another book followed, *On Human Nature* (1978),[67] which was perceived as even more inflammatory. Here he discussed, among much else, the genetic 'explanation' of human aggression, sex differences, and morals. In 1981 he published *Genes, Mind and Culture* in collaboration with a physicist, Charles Lumsden (Lumsden and Wilson 1981), and, two years later, *Promethean Fire: Reflections on the Origin of Mind* (Lumsden and Wilson 1983). This considerable output is now far outstripped in volume, if not always in quality, by the literature generated by his critics and defenders.[68]

Sociobiology: The New Synthesis received much attention from the press. It was the subject of a front-page article in the *New York Times* some weeks before it was published; a massive publicity campaign was organized by his publishers, Harvard University Press; and Wilson himself wrote an article for the *New York Times Magazine* (published 12 October 1975) entitled, though perhaps not by him, 'Human Decency Is Animal,' in which he outlined some of his ideas for a general audience.[69] He was also extensively interviewed on television and by various other magazines and newspapers. All this is reminiscent of the publicity that surrounded the publication of Jensen's article in the *Harvard Educational Review* six years earlier. It is therefore not unfair to say that Wilson must bear some responsibility for the political attacks that were made on his ideas, if not for the vehemence with which they were expressed.

One of these attacks came from a 'Sociobiology Study Group' consisting of scientists and other professional people who were members of Science for the People, a generally radical organization that disseminates important information and criticism concerning the misuse of science and technology. The study group accused Wilson of racism, sexism, and other forms of intolerance, and of having joined 'the long parade of biological determinists whose work has served to buttress the institutions of their society by exonerating [these institutions] from responsibility for social problems.' The tone employed was strident and polemical. This was later seen as a tactical error, and regretted by some members of the group. It was argued, for instance, that sociobiology simply reflects the social arrangements of the capitalist system. This could be seen from the fact that Wilson's research was funded by the state (an accusation that was as irrelevant as it was unjust, since almost all scientists are so funded), and that he worked at Harvard, an institution with close links to capitalist wealth and the centres of political power (but so did some of the members of the group that was criticizing him). These charges and innuendoes were angrily rejected by Wilson as 'self-righteous vigilantism.' Yet his own position concerning the political implications of 'popularizing' biological research has been more than a little equivocal. Responding to an article by Medawar in the *New York Review of Books* in 1977,[70] Wilson argued that 'we should distinguish between those who wish to politicise human behavioural genetics [his critics] and those who wish to depoliticise it [including, of course, himself]. Opprobrium, in my opinion, is deserved by those who politicise scientific truth, who argue the merits of analysis according to its social implications.' The irony in this statement is that Wilson was not defending himself on this occasion (at least directly), but Arthur Jensen; for it was Jensen's idea that Medawar was criticizing in the article to which Wilson was taking exception. Moreover, the argument that 'opprobrium is deserved by those who politicise scientific truth' came a little oddly from the man who had written in the *New York Times Magazine* that genetic differences between men and women suggest that 'even with identical education and equal access ... men are [always] likely to play a disproportionate role in political life, business, and science,' and then had quickly added, 'but that is only a guess.'

One of Wilson's problems with his critics is that he purports to see the issue that separates him from them purely in terms of what logicians call the Exclusive Or – either genes or culture, but not both. 'Evolution has not made culture all-powerful' (1978: 18). This 'mistake' is made by 'many of the more traditional marxists' and by 'a surprising proportion of anthropologists and sociologists.' Elsewhere he has written (1977: 136) that the social sciences are imbued with the view that 'human social life is the nearly exclusive product of cultural determinism constrained only by the most elementary and unstructured biological drives.' But it is Wilson who is mistaken. There are some cultural determinists in the social sciences, but many fewer today than in the past. Cultural determinism is far from the prevailing view; indeed in anthropology it is probably more weakly represented today than at any time in the history of the discipline. Wilson is himself a determinist, but a biological determinist.[71] Although he frequently disavows this, his denials carry little conviction and seem chiefly to be designed to throw his critics off balance. An example is his often-quoted statement that in human societies 'genes have given away most of their sovereignty' – a form of words, incidentally, that is typical of Wilson's style of scientific writing.

The statement has been used by his defenders as evidence that he has been badly misrepresented; but when it is put in context it can be seen that this is not so. He asserts, for example, that the culture of any society 'travels along one or other set of evolutionary trajectories whose full array is constrained by the genetic rules of human behaviour.' This leaves little room for cultural evolution, particularly when it is reinforced by such statements as the following: that 'the deep structure' of human nature is 'an essentially biological [that is, genetic] phenomenon' (1978: 10); that the 'picture of genetic determinism emerges most sharply when we compare selected major categories of animals with the human species' (p. 20); and that the genes 'hold culture on a leash' (p. 167). It is true that he quickly adds that 'the leash is very long,' but this is a typical equivocation and there is little doubt that he attaches almost no importance to the qualification. For example, he tells us that the behaviour to be explained by sociobiology should be 'the most general and the least rational of the human repertoire, the part furthest removed from ... the *distracting vicissitudes of culture*' (p. 35, emphasis added) – surely a most revealing remark! He adds that although cultural learning can undoubtedly increase biological fitness, there is a limit to this 'cultural mimicry'; that 'methods exist' by which cultural learning can be distinguished from 'the more structured forms of biological adaptation' (p. 33); and (perhaps most telling of all) that 'even though human behaviour is enormously more complicated and variable than that of insects, theoretically [sic] it can be [deterministically] specified' (p. 73).

It is quite false to suppose that those who reject biological determinism must then be cultural determinists. Stephen Gould is an example of the many scientists and social scientists who are not. Gould has written (1977: 252): 'The issue is not universal biology versus human uniqueness but biological potentiality versus biological determinism.' Humans cannot do those things that are forbidden by their genetic constitution. We cannot, for example, photosynthesize; nor can our bodies manufacture all the vitamins we need to survive. Our genes restrain us in many ways; but as Gould says, if that is all Wilson means by genetic control he is saying very little. Of course, as Gould knows, that is *not* all he means. He is not making a statement about human potentiality but about genetic necessity, qualify this how he will. That is why it was inevitable that he should claim that there exist specific genes for specific traits like xenophobia, homosexuality, and others we have already noted. There exists no empirical evidence to support this claim but much reasoning that contradicts it.

By the use of metaphor and analogy it is possible to say, as many biologists including Wilson do say, that certain animal behaviours that are genetically caused are 'altruistic' or 'spiteful'; that certain species of ants take other ants as 'slaves'; that some species of termites 'engage in chemical warfare' (by spraying attackers with a gluelike chemical substance); or even (as Dawkins does, 1976) that some species – for example, the red grouse of the Scottish highlands – engage in 'family planning.' As Rhinelander has said, 'Whenever we are confronted by a new phenomenon – our first impulse is to assimilate it to the models and categories to which we are accustomed ... The difficulty is, of course, that analogies and metaphors are incomplete. They mark the resemblance but they ignore the differences. Thus they may lead us to assume an essential identity merely on the strength of a vivid but accidental similarity' (1973: 24). And to do so may be dangerously misleading. The distinguished population geneticist John Maynard-Smith defended Richard Dawkins's use of the term 'the selfish gene' (the title of Dawkins 1976) by saying, 'I suppose that Dawkins referred to genes as selfish because he imagined no one would take him seriously' (1982). Would that this were so.[72] It is evident that many people, including some scientists, take their analogies very seriously; indeed their behaviour shows a tendency to drift into treating their analogies as if they were homologies. Thus the term 'altruism' is attached to certain animal behaviour that appears to resemble the regard for others that humans practice, and it is then assumed (especially by reductionists like sociobiologists) that a seemingly like effect must have a like cause – in the present case, a genetic cause. But this is to commit a logical mistake. Like effects do not necessarily have like causes.

Wilson has argued that reduction is 'the traditional instrument of scientific analysis' (1978: 13). He is partly correct. As a method, explanation by reduction has been very fruitful in science, even when, as has often been the case, *complete* reduction proved to be impossible. One of the most successful complete reductions in the history of science was Newton's reduction of Kepler's and Galileo's laws to the law of universal gravitation. But such reductions are rare. It is by no means established that the whole of chemistry will some day be reduced without remainder to the laws of physics (though it has been partly reduced by quantum physics),[73] let alone that human behaviour in all its complexity will ever be reduced without remainder to biology. Wilson is fond of saying that, when some scientist says that something is impossible, it is a sure sign that it is about to happen. But there is no evidence that a complete reduction is about to take place in sociobiology. Let us again take altruism, one of the key behavioural traits of interest to sociobiologists, as an example. According to Wilson, human altruism is ultimately explainable in purely genetic terms. One of the pathways through which this trait operates is the process of kin-selection, which is present in many animal species. Many 'altruistic' acts by animals, as Gould puts it, 'seem to defy a Darwinian explanation' (1977: 255).[74] 'On Darwinian principles, all individuals are selected to maximise their own contribution to future generations. How then can they willingly sacrifice or endanger themselves by performing acts of benefit to others?' The answer is 'charmingly simple in concept, though complex in technical detail.' By sacrificing himself to save his relatives the altruist may increase his own genetic representation in future generations.[75] The biologist J.B.S. Haldane once characteristically remarked that he would lay down his life 'for two brothers or eight cousins.' As Gould explains, 'in most sexually reproductive organisms an individual shares (on average) one-half the genes of his sibs and one-eighth the genes of his first cousins' (1977: 255). Hence, 'the Darwinian calculus' would favour Haldane's sacrifice.

This theory brilliantly explains much observed animal behaviour. Does it explain analogous human behaviour? Gould considers the classic issue of grandparent sacrifice among the Eskimo (the Inuit). This was the practice by which, when food supplies dwindled and the family was forced to move, the grandparents willingly stayed behind to die rather than risk the survival of the whole family by slowing its progress. This might seem a perfect example of genetically based altruism (to ensure the survival of the grandparents' genes). Perhaps, it is. But, 'an extremely simple, nongenetic explanation exists: there are no altruistic genes ... The sacrifice of grandparents is an adaptive, but nongenetic, cultural trait. Families with no tradition of sacrifice do not survive for many generations. In other families, sacrifice is celebrated ... Children are socialised from their earliest memories to the glory and honour [of it]' (1977: 256).

What Wilson has called 'hard core' altruism is the altruism of the honey-bee, which 'commits suicide' by stinging intruders to the hive. But there is another form of altruistic behaviour that Wilson calls 'soft core,' though it is more usually referred to by biologists as 'reciprocal,' namely altruism toward non-relatives. The notion is that the individual performing an altruistic act will do so in expectation of a similar favour from the beneficiary at a later time. This kind of behaviour is especially noted among the primates (chimpanzees have been seen to share their meat and adopt their orphans), but other animal species are known to practice it also – for example, *Melanerpes formicivorous*, the acorn woodpecker. These birds share mates and raise their young co-operatively, and some of the adult birds renounce mating altogether in order to help with the rearing of the young. The practice as it exists among animals can usually be shown to have a biological basis; but sociobiologists experience great difficulty in explaining it genetically in humans. Wilson's attempt is particularly unconvincing. Chapter 7 of *On Human Nature*, 'Altruism,' is a masterpiece of myth-spinning rationalization. The complex and elusive argument defies summary; but in essence it appears to amount to the Hobbesian-like claim that human altruistic acts that are performed for others than kin are done out of vanity or for glory. 'Lives of the most towering heroism are paid out in the expectation of great reward, not the least of which is belief in personal immortality' (p. 154). Lesser acts are selfish and calculated according to the degree of expected approbation; and they may be hypocritical (p. 156). 'The capacity for soft-core altruism can be expected to have evolved primarily by [natural] selection of individuals' (genetic 'explanation'), but it developed culturally because 'hard core altruism is the enemy of civilisation'; that is to say, kin selection favours kin over the greater good of society, and hence the 'greater harmony and homeostasis' that soft core altruism brings must inevitably override it (cultural explanation, but with a strong genetic base).

Darwin was led to believe that human evolution was predominantly biological. His nineteenth-century opponents countered this by arguments about human uniqueness (mind, reason, and the existence of the soul). It was therefore important for him to be able to show that embryonic counterparts of these apparently unique characteristics could be found in other animals, and that they had evolved, in the Darwinian manner, through a long series of small changes brought about by natural selection. As he wrote in his notebooks, if nature 'has put Man on the throne of reason she has also placed a series of animals on the road that leads up to it.' It is entirely understandable therefore that, although he was not unmindful of cultural factors and indeed believed that they became more important as civilization advanced, he failed to realize their proper significance. Biologists today, like Wilson, are under no such compulsion and cannot be so easily excused. To admit the reality and importance of exosomatic evolution is not to deny the truth of biological evolution. Why, then, does it pose such a threat to sociobiologists?

The claim made by 'exosomatic evolutionists' is that we have developed through a process of biological evolution a great flexibility in behavioural response (to use Gould's expression), which enables us to create a world of immense complexity outside ourselves. It is likely that we developed this flexibility because we evolved a brain that is much larger in relation to body size than any other animal,[76] and a relatively unspecialized body form that gives us the capacity for exploiting a range of different environments and different styles of living that far exceeds that of any other animal. We have developed language that goes much beyond the signalling and expression of emotions that we share with other creatures. Chimpanzees have recently been shown to be startlingly close to us genetically. We are capable of communicating with them, and they with us, in an elementary way (if only by reading each other's cues); but the capacity to convey information[77] is only one aspect of human language. Language is not the mere making of meaningful sounds and signals. The evolution of language in humans has enabled us (among other things) to *describe* the world, which has led to the ability to speculate and to formulate abstract thoughts, to the capacity to distinguish truth from falsity, and to argue about ideas. In these and other ways we have 'changed our nature.' It may be that some creatures have some rudimentary consciousness of self; but our consciousness of self has led us to know our own mortality, and to conceptualize the future and the past. The point is not that our uniqueness denies our links with a biological past; it is that our culturally evolved differences from other creatures have enabled us to shake off most of its constraints. Not all the effects of this evolution have been fortunate for us. Many animals exhibit aggression, but we are the only animals that deliberately make war, and this is because we are the only animals that are capable of making it. As some wit has remarked, 'it is possible that a chimpanzee, faced with the option of becoming human, would take one look at human society and turn down the offer.' The fact that we are not wholly admirable creatures may be one of the reasons for the popular appeal of sociobiology. For if we are ultimately explainable by reduction to our biological constitutions, there is nothing we can do about it, and we can rest discontented.

5

Pseudo-science

Science falsely so-called.
1 Timothy 6:20

There are certain ways of looking at the world and of approaching an understanding of it that can be called 'scientific,' and that persist no matter how the content, the practice, and even the methodologies of science may change in detail. There exists, in short, a scientific tradition and a scientific 'attitude.' There are boundaries around science of a practical as well as a philosophical kind that most scientists see it as their duty to protect. Within science, ideas have to prove their mettle. There is, as sociologists say, a 'reception system' for ideas. The scientist has to get his ideas accepted by other scientists if he is to make 'a contribution to science.' This informal but nonetheless potent process of scrutiny and criticism by peers is reinforced nowadays by the existence of formal 'gatekeepers.' Scientific referees judge research results before publication in the professional journals, and granting committees of scientists assess applications for money to support new and continuing research. The system operates primarily to decide what is to be accepted as a likely contribution to the advancement of science; a largely unintended consequence is that it operates also to determine what is scientific and what is pseudo-scientific.

An idea, theory, problem, experiment, or research result generated 'within the boundaries of science' will usually be judged scientific even if it is rejected, or ignored, or looked upon with scepticism by other scientists. There are many instances of ideas, theories, and research results that have had to wait a long time for acceptance, and sometimes they are never accepted. An important recent example of the former is the work of Peter Mitchell, a biochemist who won a Nobel Prize for chemistry in 1978 after some twenty years during which his ideas were ignored and even ridiculed. The work for which he won the prize had to do with the processes by which organisms like plants and bacteria convert light into energy. The orthodox view was that this must be done chemically. Dr Mitchell proposed a physical explanation. The fact that non-scientific or extra-scientific considerations as well as purely scientific ones may enter into the making of scientific judgments is apparent in this case. Dr Mitchell had retired from a regular university research post at the University of Edinburgh for reasons of health, and had set up his own small research establishment in a remote part of England. Hence, not only were his ideas heretical, he had become an outsider. An even more dramatic instance of long neglect is the case of the American geneticist Barbara McClintock. Although Dr McClintock's work in corn (maize) genetics, and her outstanding gifts as an experimenter, have long been recognized by her fellow molecular biologists (she was elected a member of the U.S. Academy of Sciences in 1944 at the age of forty-two, only the third woman to have been so honoured up to that time), her revolutionary discovery that genes can transpose from one chromosome to another fell on stony ground when she announced it in 1951 and remained neglected for another quarter-century until, as James Watson put it, 'science caught up with her.' In 1983, at the age of 81, she was awarded a Nobel Prize.

We see here a somewhat similar case to that of Dr Mitchell: a totally dedicated person, preferring for the most part to work alone, seen as too much of a maverick to fit in easily to the routines of university life. Both cases illustrate an inherent conservatism in science, and a scepticism about ideas that are hard to fit into the current mainstream of thought. But there is something else of great importance to be learned about science from these cases: that scientists are willing to learn from their mistakes, to be convinced that they were wrong, and to accept the principle that heretics are sometimes revolutionaries. In the words of Barbara McClintock's intellectual biographer, Evelyn Keller: 'If Barbara McClintock's story illustrates the fallibility of science, it also bears witness to the underlying health of the scientific enterprise. Her eventual vindication demonstrates the capacity of science to overcome its own characteristic kinds of myopia, reminding us that its limitations do not reinforce themselves indefinitely' (1983: 197). Dr Keller adds – which words should be graven on the brows of all contemporary sociological critics of science – that 'however severely communication between science and nature may be impeded by the preconceptions of a particular time, some channels always remain open; and, through them, nature finds ways of reasserting itself.'

The idea that accepted theories and models blind scientists to new possibilities is not new. Max Planck's half-serious suggestion that before a new idea is accepted 'all the old men have to die off' is often quoted. He was anticipated, however, by the great French chemist Lavoisier,[1] who wrote in 1795: 'I do not expect my ideas to be adopted all at once. The human mind gets creased into a way of seeing things. Those who have envisaged nature according to a certain point of view during much of their career rise only with difficulty to new ideas. It is the passage of time therefore which must confirm or destroy the opinions I have presented.' Too much can be made of this, however. Einstein was little known in 1905 when he made his revolutionary proposal about the quantum nature of light. His scientific biographer, Abraham Pais, has said: 'The physics community at large ... received the light-quantum hypothesis with disbelief and with scepticism bordering on derision ... From 1905 to 1923 Einstein was ... the only one, or almost the only one, to take the light-quantum seriously' (1982: 357). At the same time, however, Einstein's only slightly less revolutionary theory of special relativity, also advanced in 1905, was almost immediately endorsed by the majority of physicists, though this was perhaps assisted by the fact that Poincaré was on the verge of making the same discovery – that is to say, that the community was intellectually more prepared to accept special relativity than the light-quantum.

Other explanations of the resistance to new ideas have been advanced. One is that workers within a specialism may be reluctant to accept the ideas of someone whose work does not lie primarily within the specialism. This is no doubt one of the reasons, though only one, for the lack of recognition accorded to Alfred Wegener's continental drift hypothesis. This was first presented by him in 1912; but it was neglected, and frequently contemptuously dismissed, for nearly half a century. It is quite clear today from the historical evidence that geologists were not being irrational when they greeted his idea sceptically. The point here, however, is simply that Wegener had no formal credentials as a geologist. He had a doctorate in astronomy and was a professional meteorologist. Perhaps the most famous case is that of Gregor Mendel (1822–84). It is not quite true to say that, when he published the results of his famous experiments with peas in 1866, he 'achieved instant oblivion' (Bronowski 1973: 385). It is true that he was an 'obscure monk' and not a professional scientist (though that was not at all unusual at that date), but his papers were quite widely circulated and they were included in a catalogue of scientific literature published in 1879 in England. Many botanists were aware of his results. The problem was that their significance was not grasped and he had to wait for recognition until 1900, when his work was 'rediscovered,' independently, by three other workers.[2] The essential point is not his obscurity or his lack of professional status, but the fact that his ideas failed at first to cohere with the currently accepted body of knowledge and practice within the discipline.

Nor does the reception system operate solely on marginal figures. When Lord Rayleigh, an established scientist, submitted a paper to the British Association for the Advancement of Science in 1886, his name was somehow omitted and the paper was rejected. When the authorship was discovered it was at once accepted. Michael Polanyi used to make frequent reference to an incident that happened to Lord Rayleigh's son, who succeeded to the title and was himself a distinguished scientist. In 1947 he published a paper in the *Proceedings of the Royal Society* reporting some experimental results that, Polanyi claimed, could if correct have been 'far more revolutionary than the discovery of atomic fission' (1951: 12): 'Yet when this paper appeared and I asked various physicists' opinions about it, they only shrugged their shoulders. They could not find fault with the experiment, yet not one believed in its results, nor thought it even worth while to repeat it.'

There are, however, sound reasons for the exercise of scientific caution in admitting new ideas. One is the avoidance of fraud, which may bring science into disrepute and is in any case self-defeating, since it is not the aim of science to make false statements about nature. Another is the possibility of self-deception on the part of the investigator (or investigators), one of the manifestations of which is the occurrence of observational and experimental 'pseudo-effects.' Fraud and self-deception are not always easy to distinguish, as we shall see in the case of parapsychology. Indeed a nice example of their interconnection is provided by Piltdown Man, which is perhaps better considered a hoax than a fraud (though the matter is still undecided), but in any event was a fake, and a fake that deceived a number of anatomists and anthropologists of the highest standing.[3] Piltdown Man stands as an awful warning of the need in science to guard against fraud and the deceptive nature of appearances.

Enough has been said, even in this sketchy account, to illustrate the importance of the reception system in science and to indicate the extent to which scientists (to use Polanyi's term) 'watchfully resist' ideas that may be erroneous. Justice is not always done; mistakes are made; prejudice and personal animosities sometimes take a hand; received ways of thinking and doing may stand in the way; disputes over scientific claims may sometimes be accompanied by acrimonious rivalry. None of this is surprising. Scientists are human; science is a social system, and as such cannot be expected to be perfectly rational. Nevertheless, there operates in science what Bernard Barber (1961) has called a powerful norm of open-mindedness – not because scientists are more open-minded in general than other people, but because open-mindedness, highly critical open-mindedness, is necessary to the advancement of science. Yet if science is an open-minded as is claimed, why does this not extend to ideas and doctrines that lie 'outside the boundaries of science'? This is a question that is often asked by members of the public. The answer, in general (we shall look at this more closely later), is that pseudo-scientific ideas are ideas that fail to adhere to generally accepted principles of scientific discourse and scientific practice. Their rejection on this ground is not purely arbitrary but, as I shall argue, rationally defensible.

PSEUDO-SCIENCE, SCIENTISTS, AND THE PUBLIC

Thou god of our idolatry, the press.
Thou fountain at which drink the good and wise ...
William Cowper, *The Progress of Error*

Pseudo-science has a long history and has appeared in many exotic guises. Yet it is only comparatively recently that organized science (represented, for example, by associations for the advancement of science) has come to take its existence seriously. Why has this come about? First, there are many pseudo-scientific doctrines that seek public legitimation and support by claiming to be scientific; and there are others that are anti-scientific in the sense that they purport to offer alternative accounts to science or claim that they can explain what science cannot explain. Both are thought to pose a potential threat to the public understanding of science. Many people have difficulty in distinguishing science from impostures because they have only a very vague idea of what science is about. Secondly, there has been a growing tendency among the educated public and especially among intellectuals (including even some scientists) to challenge the right of scientific opinion to a deciding voice in what is to be regarded as science and what as pseudo-science. Thirdly, many scientists believe that the popularization of pseudo-scientific ideas through publishing, film, and the broadcast media helps to confuse the public about the true nature of science, fosters uncritical acceptance of false notions generally, and, in some instances, may play on superstitions and irrational fears.

Television documentaries on pseudo-scientific subjects (often using well-known 'media personalities' as narrators to increase their appeal) frequently present them either as scientifically based and 'supported by experts,' or as demonstrating the limitations of science.[4] And there has been in recent years a barrage of sensational books, out of which a great deal of money has been made for their authors and publishers, on such topics as flying saucers, the Bermuda Triangle, the 'psychic' spoon-bending of Uri Geller, pyramidology, and the secret life of plants (which are said to respond to human thoughts and emotions, especially when hooked up to a lie detector). The preposterous book by David Rorvik, *In His Image: The Cloning of a Man*, came close to being an extraordinary success until it was exposed as a fraud in the course of a successful lawsuit brought against the author by a British scientist in 1981.[5] Van Daniken's equally absurd *Chariots of the Gods* (1971) had sold over six million paperback copies by the end of 1979, and was at one time second on the *New York Times* list of non-fiction best sellers.

These sensational books are widely promoted by their publishers; the authors are sent on promotional tours and frequently appear on radio and television 'talk shows.' They also receive considerable exposure in the popular press. When they arouse controversy, perhaps by rebuttals from scientists, the publishers do their best to exploit it. The publishers of Immanuel Velikovsky's work (see later section), for example, have done this very successfully, even to the extent of publishing a posthumous reply to his critics four years after his death. A book highly critical of parapsychology (see later section) written by two New Zealand psychologists, David Marks and Richard Kammann, *The Psychology of the Psychic* (1980), was rejected by more than thirty publishers before it found a taker. As one reviewer remarked: 'Publishers know their market, and while eager for material supporting the supernatural, are reluctant to publish books that attempt to expose it.'[6]

There is, however, a more self-interested and less directly public-policy-oriented reason for disquiet among scientists: the fear that pseudo-scientific ideas may, by feeding anti-science attitudes, create an atmosphere of mistrust of science; and since in a free society the public is also the electorate, this may affect state funding of legitimate scientific research. One of the most astonishing aspects of the widespread belief in the paranormal is its attraction for persons of all levels of social class and education. For example, in 1975 Marks and Kammann (1980) surveyed one of their psychology classes (304 students) at the University of Otago and found that more than 80 per cent of their students believed in telepathy and 30 per cent actually believed they possessed psychic powers. A survey of the readership of *New Scientist* in 1973 revealed that 67 per cent of the respondents held extrasensory perception to be either an established fact or a likely possibility. Fifteen hundred people responded to the questionnaire. Only 3 per cent of the sample believed ESP to be impossible. (*New Scientist* was then publishing about 70,000 copies a week.) Of the respondents, 63 per cent had university bachelor's degrees and 29 per cent had higher degrees as well. The social climate prevailing at the time these surveys were made should not be overlooked; nor should the power of the popular media to manipulate opinion with sensational stories; but these are startling results nevertheless.

Scientific opinion is often divided as to how far scientists should go in attacking pseudo-scientific ideas lest their criticisms create ill will. There have been, and are, many such ideas that are simply cranky and that very few people believe: almost no one now believes that the earth is flat, or that it is hollow, but there are still some flat-earthers who argue that the photographs taken from outer space are faked, and there is said to be a hollow-earth cult in Germany. But these are of little concern. Two types of pseudo-science do trouble scientists: fully worked out *systems* of pseudo-scientific ideas like parapsychology or astrology, and less fully developed ideas that make a direct attack on scientific principles, such as those of Immanuel Velikovsky.

THE CASE OF IMMANUEL VELIKOVSKY

But one must not be misled by the evidence.
Freud, *Moses and Monotheism*

In January 1950 *Harper's* magazine published an article by one of its staff writers, Eric Larrabee, entitled 'The Day the Sun Stood Still.' The article contained an outline of the ideas contained in an unpublished manuscript written by a physician and psychoanalyst, Immanuel Velikovsky. This was the beginning of an extraordinary confrontation between science and pseudo-science that lasted for more than a quarter of a century. Indeed, it is still not finally settled, though Velikovsky died in 1979, and a posthumous book by Velikovsky giving an account of the controversy from his side was published in 1983. (For a full, highly partisan account of the early history of the affair see Juergens 1966.)

In fact there were two manuscripts, which Velikovsky had been trying for several years to have published. One of them, called *Worlds in Collision*, had been read in 1946 by the science editor of the *New York Herald Tribune* and he had written a laudatory piece about it in the newspaper in August of that year. Little attention seems to have been paid to this at the time, in spite of the fact that the science editor had described it as 'a stupendous panorama of terrestrial and human histories which will stand as a challenge to scientists.'

According to Juergens the reason for the long delay in finding a publisher was that 'the heavily annotated text was too scholarly for the book trade' (1966: 19), a remark that may strike the reader, in the light of what follows, as somewhat bizarre. Eventually, however, the trade books editor of the Macmillan Company was persuaded to take it. It was at this point that the *Harper's* article appeared, soon to be followed by more coverage in *Reader's Digest* and *Collier's* magazine. By this time science, in the guise of the distinguished American astronomer Harlow Shapley, had begun to stir. It was clear to Shapley that some very distinguished, if misguided, persons had by now read the manuscript and had been impressed by it, and that these included some scientists, notably Gordon Atwater, the curator of the Hayden Planetarium in New York City.

There are three responses that scientists can make to what they regard as pseudo-science. The first is to ignore it, the second is to answer it with scientific argument, and the third is to attempt to suppress it. Shapley (wrongly, as most concerned scientists now think) chose the third option. He began to bring pressure on Macmillan to reverse the decision to take the book (Macmillan was a major publisher of scientific textbooks and was thus particularly vulnerable to scientific opinion) and enlisted the assistance of a number of colleagues in various disciplines including geology, archaeology, anthropology, and oriental history to publish denunciations of Velikovsky's ideas. His former graduate student (but by then a well-known astronomer in her own right) Cecilia Payne-Gaposhkin was particularly active in the campaign. Much has been made by Velikovsky's supporters of the fact that these critics admitted that they had not read the work itself, but only what had appeared in the magazine articles.

When the book appeared it was an instant popular success; but by now more scientists were mobilized and Macmillan was forced, in view of the mounting volume of protest, to make over the publication rights to the Doubleday Company, which had no scientific textbook department and was willing to take it. It may well have been the fact that Macmillan was a major scientific publisher that most alarmed Shapley and his associates, since the publication of the book under the Macmillan imprint might have been taken as a sign of scientific approval. A British edition quickly followed, published by Gollancz by arrangement with Doubleday, and received the same hostile reception from the scientific community. Doubleday continued to publish Velikovsky, including the sequels to *Worlds in Collision* called *Ages in Chaos* and *Earth in Upheaval*, and a book published two years before his death, *Mankind in Amnesia*. It was also the publisher of a series of essays defending his ideas, some by Velikovsky himself, under the title *Velikovsky Reconsidered*. *Worlds in Collision* went through seventy-two printings in its first twenty-five years of life and is still selling.

The campaign against Velikovsky was well-orchestrated, but it failed in its main purpose, which was to have the book suppressed. Most people now agree that it represented an over-reaction and was in many ways a discreditable episode, not least in that it led to the firing of an editor who had been with the firm of Macmillan for more than twenty years, and also of Gordon Atwater who, apparently unaware of the hornets' nest on which he was sitting, had proposed a public display at the Hayden Planetarium illustrating Velikovsky's ideas and was obdurate when pressed to abandon his plans. There is also little doubt that the publicity aroused by Shapley's campaign materially assisted sales of the book, producing the very result that Shapley wished to avoid. What, then, was all the fuss about, and what was it about Velikovsky's ideas that generated so much heat?

Velikovsky had moved to the United States in 1939 from Palestine, where he had practised as a physician, in order to write a book about Freud. In the course of his research he was led to examine events in the life of Moses and thence to the idea that the catastrophes and apparently miraculous events (such as the parting of the Red Sea) recorded in the Old Testament might have had some basis in fact. This led him, in turn, to study the records of catastrophes found in the legends of other societies, and he was struck as many before him had been by the apparent similarities between them. The next step, taken over a period of several years, was to the notion that, since seemingly related events appear in the records of societies that were widely separated geographically, the events described could not be accounted for as merely local natural phenomena but must have had some more universal or cosmic cause. His solution to this problem was that the cause must be traced to major disturbances in the solar system, including the ejection of Venus in the form of a comet from the body of the planet Jupiter, the close approach of this comet and also of the planet Mars to Earth, and similar titanic upheavals. It follows from this sequence of 'discoveries' that all these things must have happened no more than a few thousand years ago and with relatively short intervals of time separating each one from the others. This staggering conclusion – that the present state of the solar system is of recent origin (perhaps of no more than about three thousand years) – is the central theme of Velikovsky's work, not merely of *Worlds in Collision*, but of all his subsequent writings as well.

It is hardly surprising, therefore, that the scientific community was at once contemptuous, and yet alarmed about the possible effects that these ideas might have on the public; for if Velikovsky were right, the prevailing conception of the solar system – that it must have existed in its present general configuration for several billion years – would have to be abandoned, and most of modern geology, evolutionary biology, and physics would have to be revised and rethought. This, indeed, is what Velikovsky himself believed, and what his supporters still believe. It is important, however, to consider the manner in which Velikovsky arrived at these conclusions. He began from the assumption, perhaps always the certainty, that the accounts of the happenings recorded in ancient myths and legends were, if not literal descriptions of actual historical events, then reflections or interpretations or collective memories of actual historical events. Where there were inconsistencies in the records these, he thought, as befitted a faithful student of Jungian ideas, could be ascribed to collective amnesia, the social forgetting of traumatic experiences. Then, having made the assumption that the legends in some way referred to actual events, Velikovsky asked himself, not how the events could be explained in a manner that was consistent with the existing state of scientific knowledge, but rather, how scientific knowledge must be revised to make it consistent with the events.

Some flavour of the true provenance of Velikovsky's thought may be gleaned from a curiously juxtaposed, and almost offhand, remark made in a paper written in 1972, more than twenty years after the first publication of *Worlds in Collision*: 'Some authorities ... claim that the scars on the face of the moon are older than 4 1/2 billion years. The lunar landings will provide the answer [he later claimed that they had]. Was the face of the moon as we see it carved over four and a half billion years ago, or, as I believe, less than 3000 years ago? If this unorthodox view is substantiated, it will bear greatly not only on many fields of science, *but also on the phenomenon of repression of racial memories, with all the implications as to man's irrational behaviour*' (1976: 217; emphasis added). This and other evidence[7] suggests that Velikovsky was more concerned to produce a psychoanalytic account of human irrationality and to revise Jungian ideas about the collective unconscious than he was about the truth of existing scientific knowledge or, even, about the accuracy of existing accounts of ancient history (on which he has also been challenged by experts): 'If cosmic upheavals occurred in the historical past, why does not the human race remember them? ... The task I had to accomplish was not unlike that faced by a psychoanalyst who ... reconstructs a forgotten traumatic experience' (Preface to the first edition of *Worlds in Collision*). It apparently did not occur to Velikovsky that the reason we do not remember them might be because they did not happen.

The references to scientific matters that infuse his work are a curious mixture of acuteness and naivety. Some of his astrophysical 'predictions' (on which more will be said later) that he used to support his assertion of recent celestial catastrophe on a grand scale were remarkably astute and were later confirmed. However, he held, for example, that the solar system is 'actually built like an atom,' and asserted that the planets could jump, as electrons in the atom do, from one orbit to another – only on a time scale of 'hundreds or thousands of years' instead of 'many times a second' (1967: 387).

Velikovsky's supporters were drawn from a wide range of professions and employments. There were engineers, philosophers, scholars in the humanities, social scientists, journalists of course, students (especially in the years of cosmic upheaval on the campuses), and others on both sides of the Atlantic unidentified by occupation who wrote to the newspapers, to semi-scientific journals like the *New Scientist*, and to highbrow 'humanist' publications such as the *Times Literary Supplement* and the *New York Review of Books*, whose reviewers had criticized Velikovsky's ideas. Societies were formed, for example the Institute for Inter-Disciplinary Studies in Britain, and the Student Academic Freedom Forum in Portland, Oregon. The two sides to the dispute might be characterized as belonging to different species, since nothing resulted from their intercourse. Pro-Velikovskians accused the opposition of misreading the sacred texts. Scientists, who were in any case unimpressed by the texts, responded with findings that refuted Velikovskian claims. Rejoinders followed from his supporters, and sometimes from Velikovsky himself, citing other scientific findings that appeared to contradict those first used by the scientists. The scientists would reply that the second findings were not inconsistent with the first. And so on. Each side tended to misunderstand the other, sometimes, it seems, deliberately.

One of the objections frequently made (and still made) by his supporters is that he was continually denied the right of reply in the scientific journals in which he was denounced. This, though true, is not a particularly compelling objection since Velikovsky had chosen to publish his ideas in a public forum in the first place. However that may be, it is a fact that no official scientific body saw fit to subject his ideas to measured critical analysis until 1974, when the American Association for the Advancement of Science was persuaded to do so. Some of the papers offered at a symposium organized at the annual meeting of the association in San Francisco in that year have been reprinted in book form (Goldsmith 1977). Velikovsky's supporters (who were well represented at the AAAS meeting) organized their own conference a few months later at McMaster University in Hamilton, Ontario. Velikovsky spoke at length at both these conferences.

One of the most detailed scientific critiques of Velikovksy at the AAAS meeting was that provided by Carl Sagan. He was subsequently attacked by Velikovskians as biased and unfair; but a careful reading of his paper (Goldsmith 1977: 41–104 and Sagan 1979: 81–127) shows this to be unjustified, although some of his facts and interpretations have been challenged. The crucial issue, however, is whether or not scientists should spend time criticizing and exposing pseudo-scientific ideas. Many are reluctant to take the time to do so. Others doubt its wisdom and some have argued that the AAAS symposium was a blunder. On this Sagan has the following to say:

> There are many cases where the belief system is so absurd that scientists dismiss it instantly but never commit their arguments to print. I believe this is a mistake. Science, especially today, depends on public support. Because most people have, unfortunately, a very inadequate knowledge of science and technology, intelligent decision-making on scientific issues is difficult. Some pseudo-science is a profitable enterprise, and there are proponents who not only are strongly identified with the issue in question but also make large amounts of money from it. They are willing to commit major resources to defending their contentions. Some scientists seem unwilling to engage in public confrontations ... because of the effort involved and the possibility that they will be perceived to lose a public debate. (1979: 59)

Velikovsky's supporters claim that he was a scientist, though a scientist working in many different fields, a 'true interdisciplinarian.' Velikovsky appears to have shared that opinion. But if he is to be regarded as a scientist then it is entirely proper to look critically at his methodology. His ideas cannot be rejected as unscientific merely because they appear implausible to scientists, since (as we have seen) many revolutionary ideas have appeared implausible to scientists at the time they were proposed. This argument has been used, quite justifiably, by Velikovsky's supporters. Nor can it be said that an idea is unscientific for no better reason than that scientists say it is. Something more convincing is required than mere demarcation by fiat. However, public acclamation cannot settle the matter either, and in the case of Velikovsky his supporters quite clearly appealed to the public. 'The findings of science cannot be reversed by majority vote' (Asimov 1977: 11).

Much has been made of the fact that some of Velikovsky's predictions – for example, that the surface temperature of Venus is much greater than scientists believed, that Jupiter would be found to emit radio waves, that the magnetosphere of the earth extends much farther into space than was currently believed, and that moon rocks would be discovered containing remanent magnetism – have turned out to be correct, chiefly as a result of information obtained from American and Soviet exploration of space. This is intriguing because all were instances of entirely unexpected phenomena, and in science surprising results are often considered to provide particularly strong support for the theory that predicts them. There is also the puzzling discovery that Venus rotates on its axis in the opposite direction from that of other planets in the solar system, which certainly indicates that there is something special about it. Nevertheless, Velikovsky also predicted many things that were subsequently found to be false – for example, that Mars and Venus emit more heat than they receive from the sun, that the atmosphere of Venus would be rich in hydrocarbons, and that the moon's craters were formed by the bubbling of its surface when it was in a molten state.[8] More important, however, is the lack of any valid connection here between theory and prediction. A scientific prediction is a statement that can be logically deduced from a theory and a specified set of initial conditions. In Velikovsky's work we find no theory explaining how it could come about that Venus was ejected from the body of Jupiter, or how it could have caused the effects claimed. The correct 'predictions' are simply used retrodictively as evidence that Venus was so ejected and that this did have the effects claimed.

It is instructive, moreover, to look at Velikovsky's ideas about the nature of theory and fact. In the preface to the first edition of *Worlds in Collision* he says: 'If, occasionally, historical evidence [in this book] does not square with formulated laws, it should be remembered *that a law is but a deduction from experience and experiment, and therefore laws must conform with historical facts, not facts with laws*' (1967: 11; emphasis added). This is a serious misreading of the nature of a scientific law. If laws or theories were formulated only in accordance with a given set of facts (which is clearly what 'deduced from experience and experiment' means here), there would be no way of testing them independently: they would simply be ad hoc. Laws must give way to facts (that is, must be falsifiable by facts), but they are not constructed from facts. Secondly, even if Velikovsky had deduced his predictions from a properly constituted theory, it is always possible to deduce true predictions from false theories. This follows from the rules of deductive logic. That is why scientists can never prove that their theories are true; and it helps to explain why they hold all theories only tentatively and subject to replacement, even those which have been corroborated by many successful predictions. But Velikovsky's repeated assertion was that his successful predictions, or, as he preferred to say, 'advance claims,' *proved* that his ideas must be true. The reference to 'conforming with historical facts' is also significant, for it emphasizes Velikovsky's quite incorrect notion that astronomy is a branch of natural history. He was at one with his supporter Livio Stecchini, who wrote in defence of Velikovsky: 'Hence the astronomer who wants to pronounce himself today on the mechanics of the solar system cannot ignore the historical documentation and must depend on the results of historical scholarship' (1966: 137).

Velikovsky repeatedly asserted that scientific opinion was divided on many questions concerning the solar system. This, though true, was clearly intended to lead his audience to conclude that, since 'science does not know everything,' his ideas were probably correct. This is a typical pseudo-scientistic strategem. There is, of course, nothing strange or scandalous about divisions of opinion among scientists. This is a condition for scientific progress. But it is difficult to persuade the public that this is so. Velikovsky and his followers also propagated the notion that science is wedded to 'a static Newtonian conception' of the solar system – that is, to the view that it has remained essentially unchanged since its creation. This is false. It is recognized that the solar system must have had a history, and it is acknowledged that this 'history' is still going on (the sun is losing mass, the earth is slowing down, and so forth). Moreover, terrestrial catastrophes caused by external (celestial) events have certainly occurred in the past, even in the recent past – for example, the fall of meteoroids like that which is believed to have been responsible for enormous devastation in a region of the Soviet Union of 1908. But all the cosmological evidence suggests that the possibility of a series of cataclysms of Velikovskian proportions occurring so recently and within so short a space of geological time is so unlikely as to be absurd.

Velikovsky achieved considerable fame, and for the rest of his life remained an active public figure surrounded by often bitter controversy, during all of which he conducted himself with dignity and self-assurance. There is also little doubt that he was sincere in the pursuit of his ideas. But as Martin Gardner has said, 'one could say the same thing about ... Francis Gall, the father of phrenology. All the great pseudo-scientists of the past who won a popular following were sincere and mistaken. Indeed, sincerity is the main attribute that distinguishes a pseudo-scientist from a mountebank' (1981: 388). A Canadian geophysicist (York 1981) has suggested that Velikovsky was himself a mythmaker who constructed a dream, and ever afterwards pursued every scrap of information that might support it, ignoring the wealth of information that conflicted with his ideas; that out of the legends of the past he created is own legend. It should be added that in his later years he contrived a more personal legend also: the legend that he, Velikovsky, stood alone like an avenging prophet against the entire world of 'established' science, and that the reaction of his enemies was solely due to their fear of the truth that he proclaimed.[9]

PARAPSYCHOLOGY

Things and actions are what they are ...
Why then should we desire to be deceived?
Bishop Butler, *Fifteen Sermons*

Parapsychology is cautiously defined by the Oxford English Dictionary as the study of mental phenomena 'which lie outside the sphere of orthodox psychology.' This definition avoids questioning whether there are, in fact, any mental phenomena outside the sphere of orthodox psychology to be studied, and whether, if there are, they are explainable or unexplainable according to the accepted principles of science. Yet it is precisely these issues that lie at the heart of the long-standing controversy about parapsychology.

The origins of parapsychology lie in the 'psychical research' that emerged in the last quarter of the nineteenth century as a study of the claims of spiritualism. Psychical research was concerned with the investigation of the authenticity of mediums and of their ability to communicate with the dead, with the existence of ghosts and poltergeists, and with the trappings of spirit communication such as automatic writing, levitation, and ectoplasmic materialization. Although it claimed to be scientific, and attracted a number of prominent scientists such as Alfred Russell Wallace (the co-discoverer, with Darwin, of the principle of natural selection), the physical chemist Sir William Crookes (a pioneer in the study of radioactive substances and inventor of the Crookes tube, used in research on cathode rays), J.J. Thompson (generally regarded as the discoverer of the electron), and the Russian chemist Butlerov, an associate of Mendeleev at the University of St Petersburg, psychical research was frequently undertaken or supported by those already committed to a belief in the supernatural. Moreover, much of the 'research' was not scientific at all, but consisted in the quasi-legal interrogation of witnesses, the scrutiny of their written depositions, the assessment of second-hand accounts of alleged paranormal occurrences, and on-the-spot observation by persons who often had no training in the weighing of scientific evidence. The association of scientists with psychical research and especially those who were believers did, however, give it a certain standing among members of the public. The Society for Psychical Research, for example, which was established in 1882 by a group of Cambridge dons and their friends and still exists, has numbered among its presidents about a dozen Fellows of the Royal Society, although none in recent years.

The attempt to make psychical research respectable – to remove the stigma of amateurishness, to divest it of its obsessions with the supernatural in the spiritualist sense and with the religious question of life after death, and to make it more strictly 'scientific' – began in the 1920s. Parapsychology, as the study of the paranormal, was by then coming to be known, was established as a subject for systematic experimental research in 1927, at Duke University in North Carolina, under the aegis of the social psychologist William McDougall. McDougall was interested in psychic phenomena and also in eugenics and the inheritance of acquired characteristics, which he studied for many years. The work was, however, initiated by McDougall's associate, J.B. Rhine. Rhine had been drawn to spiritualism as a young man after hearing a lecture by Sir Arthur Conan Doyle, who was a prominent member of the Society for Psychical Research.[10] Under Rhine's direction Duke quickly became the centre for the study of extrasensory perception (ESP), a term coined by Rhine to include telepathy, clairvoyance, and precognition. He later included in his interests telekinesis, the supposed mental power to move or alter physical objects.

A long struggle for recognition followed. Finding it hard to gain acceptance in the established psychological journals, most of the early papers reporting the results of parapsychological experiments appeared in the proceedings of the psychical societies. Since parapsychologists aimed at scientific status rather than at the investigation of the claims of spirit mediums and of haunted houses, this situation was not wholly satisfactory to them, and in 1937 the *Journal of Parapsychology* was established under the joint editorship of McDougall and Rhine.

The possession of a journal, according to followers of Thomas Kuhn, is one of the signs of an 'emerging' scientific discipline, a step towards the production of a legitimating 'paradigm.' Nevertheless, full acceptance by the scientific community continued to be withheld; and, although the number of parapsychologists grew as other university departments entered the field, it was not until 1969 – after forty years in the wilderness – that parapsychology was recognized formally by the American Association for the Advancement of Science. This came about largely at the urging of the anthropologist Margaret Mead, who persuaded the Council of the AAAS to admit the Parapsychological Association (founded in 1957)[11] as an affiliate member. Those who now regret this move and seek to excuse it point out that 1969 was a permissive time, a time of protest on (and off) the university campuses, and of openly expressed hostility to 'established' science. It was also, as we have seen (chapter 4), the year in which Arthur Jensen's controversial paper on the hereditability of intelligence, with its supposed racist implications, appeared in the *Harvard Educational Review*. Whatever the reasons for the decision – and it is likely that political factors played a role in it – it precipitated, ten years later, an extraordinary incident.

In 1979, John Archibald Wheeler (see chapter 1) was invited to take part in a panel on 'Science and Consciousness' organized as part of the annual convention of the AAAS in Houston, Texas. At the last minute, he discovered that he was scheduled to appear on the platform with three parapsychologists. He delivered his paper in person, but later launched a sharp public attack on parapsychology. At the same time, he sent a letter to the president of the association asking for a committee to review the decision to grant affiliate status to the 300-member Parapsychological Association, which he had opposed in 1969. He requested that the committee report on the advantages, especially in fund raising, that had accrued to parapsychologists as a result of recognition, and on its effect on the public image of the AAAS. Wheeler did not mince his words. His letter stated: 'We have enough charlatanism in this country today without needing a scientific organisation to prostitute itself to it. The AAAS has to make up its mind whether it is seeking popularity or whether it is strictly a scientific organisation.' A curious history lies behind this event.[12]

What troubled Wheeler most was not that parapsychology had a long history of suspected (and in some instances proven) fraud and self-deception, but that it had recently been invaded by a number of people trained in quantum physics. The chief reason for this invasion lies in a famous paradox proposed by Einstein, Podolsky, and Rosen in 1935. The EPR paradox suggests that quantum information should, theoretically, be transferred instantaneously from one part of the universe to any other part, no matter how remote: in brief, that action at a distance is in principle possible. Einstein hoped that the paradox could be resolved and proposed a solution; but recent theoretical and experimental work indicates that the solution is inadequate. This puzzling consequence of quantum mechanical theory has been seized upon by the 'paraphysicists' (as they are inappropriately called) as a physical confirmation of the possibility of instantaneous thought transference and as a ground for explaining ESP scientifically. Two of the parapsychologists who appeared on the platform with Wheeler (Russell Targ and Harold Puthoff) were the authors of a successful mass-market book called *Mind-Reach*, published in 1979. By the time the AAAS symposium was held, 'QM-psi' was very much in vogue and had even been the subject of an international conference in Geneva in 1974. Wheeler's view was, and remains, that the attempt to link the EPR paradox to a 'quantum interconnectedness' between individual consciousnesses is baseless mysticism. This position, however, has not prevented paraphysicists from frequently citing him in support of their ideas. These ideas have not, on the whole, commended themselves to 'orthodox' parapsychologists, who have been inclined to resent the arrival of interlopers from another discipline, since they regard psychic phenomena (as did Rhine himself) as unexplainable in ordinary scientific terms. They are, anyway, less interested in explaining psychic phenomena than in demonstrating that they exist.

Dr Wheeler's attack brought a pained response from four scientists, in a letter to the *New York Review of Books* on 26 June 1980. The authors described themselves as 'physicists who have for some years been engaged in research on a possible connection between quantum mechanics and parapsychology.' One of them was Brian Josephson, of the Cavendish Laboratory, who won a Nobel Prize for physics in 1973 for his work on the theoretical physics of the solid state. Their chief objection was to Wheeler's contention that parapsychology has yet to produce a single 'battle-tested' result, whereas 'every science that is a science' has produced 'hundreds.' Their response was that parapsychology is a new science, and that even in the old sciences there are some parts that cannot meet this criterion: for example, general relativity, 'where there are only three or four hard confirmations.' The first point conveniently overlooks the fact that, by 1980, parapsychology was over half a century old, much older than, for example, phage biology, and in any event hardly 'new.' The second statement misrepresents the situation. It is true that only three or four first-order *predictions* were entailed by general relativity theory – although there have been many second-order predictions, such as, most recently, the existence of black holes – but these three or four predictions have all been corroborated many times and, in one case, the displacement of light rays in a powerful gravitational field, repeatedly confirmed by a variety of different types of experiment.[13] In contrast, there have been, over the years, numerous attempts by psychologists to replicate parapsychological experiments and to obtain similar results and all have failed.

The authors of the letter then went on to cite four cases of what they called 'competent and careful investigation' in parapsychology. The first was an investigation by Crussard and Bouvaist (reported in *Revue de Metallurgie*, February 1978) of the claims of the 'French medium' (actually a former conjuror) J.P. Girard, who is alleged to have produced large changes in the properties of metal bars, as well as to have bent them without touching them. Though there is no conclusive evience that he performed the former feats (for example, increasing the hardness of an aluminum bar by about 10 per cent) by substituting one bar for another while momentarily unobserved (a common enough type of magicians' trick), he was unable to perform his psychokinetic bending when subjected to a series of tests under controlled conditions. The second case was one of a series of experiments on the metal-bending powers of schoolchildren by Professor John Hasted of the Department of Physics at Birkbeck College in the University of London.[14] Although the particular experiment singled out by the four authors seems to have been done after Dr Hasted had become more aware of the need for proper controls, some of his earlier experiments with child subjects were carried out in the children's own homes where there was ample opportunity to cheat. It is clear that his capacity for self-deception was very great: he was, for example, at one time a firm believer in the psychokinetic powers of the best-known metal-bender, Uri Geller. Geller has now been thoroughly exposed as a highly persuasive and charismatic magician who has been on numerous occasions unable to reproduce his effects when closely watched by people who knew what they were doing.

The third example of 'careful and competent work' cited by Dr Josephson and his colleagues was an investigation by Helmut Schmidt (another Geller-believer) of 'the influence of selected subjects on the output of a random number generator based on radio-active decay' (the words of the four authors). In 'rigorously controlled experiments,' we are told, Schmidt found two subjects who could, by an effort of will, cause the generator to act non-randomly. It was Schmidt who 'discovered' in 1970 that cockroaches may have psychokinetic abilities since they turned on a randomizer that gave them pleasurable electric shocks more frequently than the laws of chance would allow (Gardner 1981: 143).

The last case cited by Dr Josephson and his colleagues was an experiment in remote viewing. The principal investigators were Targ and Puthoff. The experiment is supposed to have demonstrated the ability of 'psychic subjects' to draw accurate sketches of distant sites such as golf courses and shopping centres while these were being visited by observers who would tour the site and concentrate on all they saw. The observers were unaware of the site that had been selected before they were taken there. One of the principal investigators remained with the 'psychic subject' in a shielded room at the Stanford Research Institute, and he too was unaware of the nature of the site selected so that he could not provide the subject with cues. The subject's sketches and descriptions of the sites were afterwards ranked for accuracy by independent judges after they too had visited them. These experiments produced results that seemed to point strongly to the subjects' possession of psychic powers. This was all the more remarkable since several of the subjects had never before been tested for psychic abilities and made no claim to possess them. This fact, and the assertion by Targ and Puthoff that anyone can be a successful remote viewer (that is, that we are all psychics if we choose to be), led David Marks and Richard Kammann, two psychologists at the University of Otago, to replicate the experiments. Using exactly the same techniques, they obtained completely opposite – that is, negative – results, and their examination of the reasons for this, including the scrutiny of such of the transcripts of the original experiments as Targ and Puthoff were willing to make available to them, led to a convincing demonstration that the judges were the unwitting victims of self-deception (or as Marks and Kammann prefer to say, 'subjective validation'). They say, 'the fact is that any target can be matched with any description to some degree,' and that in comparing the description of a site made by a 'psychic subject' with his own impressions of the site, a judge will 'tend to notice the matching elements and ignore the non-matching elements' (Marks and Kammann 1980: 12–15 and 24).

Supporters of parapsychology argue that its critics can tell them how the results could have been obtained by fraud, but are unable to show conclusively that they were so obtained in any specific case.[15] This is not always true. Since most 'respectable' parapsychological research is done, like scientific research, in a closed community of practitioners without the presence of outside observers, and since most parapsychologists accept the reality of psychic phenomena, evidence of fraud will usually be circumstantial. Nevertheless there are some well-attested cases, among them that of Dr S.G. Soal, the British counterpart of J.B. Rhine, though this took many years to establish conclusively. Soal had a reputation it seems, at least in some quarters,[16] as a careful and somewhat sceptical investigator, but it was long rumoured that some startling results he had obtained in the early 1940s with a 'psychic subject,' Basil Shackleton, had been doctored, and that he had been seen by an assistant (a Mrs Albert) altering figures. When, in 1960, the British psychologist C.E.M. Hansel asked to see Soal's raw data he was told that they had been lost, though Soal had written in 1954 that the records were freely available (Gardner 1981: 219). In that year Dr Hansel published some damning evidence, but the matter was still disputed until 1978 (Soal died in 1975), when Betty Marwick, a statistician generally sympathetic to parapsychology, published conclusive evidence that Soal had cheated (Marwick 1978). Even those least willing to believe this were now forced to submit, and all of Soal's work became suspect.

Instances of fraud are, of course, to be found in science also.[17] One of the more notorious of these in recent times was that of William Summerlin. Summerlin was a 35-year-old immunologist working at the Sloan Kettering Institute in New York who confessed in March 1974 to the truth of allegations by some of his colleagues that he had faked skin grafts on two laboratory mice. He also confessed to making a false claim that he had successfully grafted human corneas into the eyes of rabbits, and there is circumstantial evidence that there may have been other incidents as well.[18] A somewhat similar type of case occurred at Rhine's laboratory where the 26-year-old director, Walter Levy, was working on the 'psychokinetic abilities' of small laboratory animals and fertilized chickens' eggs. Levy was discovered by one of his co-workers obtaining false results by tampering with a randomizing device that administered pleasurable electric shocks to rats.[19] He was dismissed by Dr Rhine in 1974. The parallel between the two cases lies not, of course, in the importance of the work (Summerlin's was potentially of great importance for the success of human transplants), but in the fact that both were young and were under enormous pressure to get results.

Mutual recriminations about the existence of fraud in science and fraud in parapsychology are unproductive. Practitioners of parapsychology in universities, unlike self-styled exponents of the occult such as Uri Geller who appear on public platforms and knowingly achieve their results by trickery, are as likely to be as embarrassed by instances of fraud as are scientists in their profession. Self-deception is another matter. Since most parapsychologists are convinced of the existence of paranormal faculties and most of their experimental 'subjects' tend to be easily convinced that they possess them, there will be strong subconscious pressures to overlook anomalous findings and to 'make the facts fit.' It is part of the discipline of a science, however, to be particularly wary of the possibility of self-deception in interpreting experimental results.

As we have seen, anomalies in science are often set aside as due to some unknown cause; in other instances (for example, Lord Rayleigh's experimental finding mentioned earlier in this chapter) no one except the 'discoverer' believes them. Moreover, there seem to be at least three different types of possibly anomalous effects (or pseudo-effects) that occur in scientific research. There are, first, those associated with the search for events predicted by accepted, and often major, theories. Examples are the claims by Joseph Weber in 1969 to have detected gravitational waves (predicted by the general theory of relativity) and a claim by P.B. Price and others in 1975 to have confirmed the existence of the magnetic monopole (predicted by Dirac in 1931). Obviously, such claims will be taken very seriously. In the two cases mentioned the search goes on. Secondly, there are experimental or observational results that purport to falsify important theories. An example is an experiment by Kantor published in 1962 that he claimed falsified the postulate on which the special theory of relativity rests, namely, the constancy of the velocity of light. This constant is much too important to be given up in the absence of overwhelmingly compelling evidence. In this case two things followed: a series of experiments each using a different technique, which produced inconclusive results, and some attempts to reconcile Kantor's result with existing theory by giving the result a different interpretation. When an exact replication of Kantor's experiment produced a negative result for Kantor and a confirmation of Einstein the matter was dropped (Magyar 1977).

Lastly, there are effects that are new and surprising and may have been found serendipitously, that is to say, accidentally and perhaps while looking for something else. The most famous cases are the alleged discovery of N-rays at the beginning of this century, an amazing instance of mass self-deception that is probably unique in the annals of science, and that of anomalous water ('polywater') by a Russian scientist, Boris Deryagin. I shall take the second case first.

Water is an astonishing substance – quite aside from its existence in both 'heavy' and 'light' forms – as Dr Felix Franks makes very clear in his brilliant account of the 'polywater' episode, which ran its course from approximately 1962 until 1972 (Franks 1981). It has many odd physical and chemical properties on which the very existence of life on earth depends. It also exhibits peculiar behaviour when it exists as small droplets, in contact with surfaces, and in confined spaces. The original 'discoverer' of anomalous water was a Soviet chemist, Nikolai Fedyakin, who was investigating the behaviour of water sealed in capillary tubes made of glass. He found that, under such circumstances, the water sometimes divided into two discrete parts, and that the upper part, which seemed to grow in volume at the expense of the lower part, was denser and more stable than the lower part. The obvious objection, of course, was that the water was not pure, as supposed, but was being contaminated in some unknown way. The work was then taken over by Deryagin, a physical chemist internationally known for his work on surface forces and respected for his meticulous experimental technique, at the Institute of Physical Chemistry in Moscow. There followed an intensive period of study and the publication of ten papers from Deryagin's laboratory between 1962 and 1966.

Franks describes how carefully the experiments were devised and the extraordinary lengths to which Deryagin and his colleagues went in order to eliminate the possibility of contamination. Nevertheless, the findings were greeted with scepticism and reserve. Then two crucial events took place, both in 1969. Scientists at the United States Bureau of Standards subjected anomalous water to infrared spectroscopy and reported that the substance that separated from 'ordinary' water was indeed water with a different molecular structure – polymerized water (hence 'polywater'). A flurry of activity then occurred and by 1972 some four hundred scientists had published results with this exciting new substance. The second event was the publication in the normally cautious journal *Nature* of a letter from an American physical chemist warning that this polymer was 'the most dangerous substance on earth' and that it should be treated in the same way as 'the most deadly virus' until its safety was established. The mass media were now fully alerted, and the result was a flood of often alarmist literature for public consumption, frequently drawing a parallel between polywater and Kurt Vonnegut's famous imaginary substance 'ice-nine,' a form of ice with a melting point high enough eventually to freeze all the water on the earth's surface. For obvious reasons, the military now entered the arena and much research, though unclassified, became defence-funded.

In many cases pseudo-effects, or supposed pseudo-effects, are small and may be almost imperceptible. In this case it was not so much that the effects were small, but that the polywater seemed exceedingly hard to produce (it appeared in only about half of Deryagin's capillaries) and was available for testing in only minute quantities (never more than a few micrograms, according to Franks). Thus, very advanced instrumentation and highly sophisticated techniques had to be used if plausible results were to be obtained either supporting or refuting the existence of the supposed new substance. In the event the sceptics won the day; Deryagin, who had hitherto rebutted failures to replicate his findings on the ground that they were caused by imperfect experimental techniques, reported in *Nature* (vol. 232, p. 131) that the anomalous properties 'should be attributed to impurities rather than to the existence of polymeric water molecules.'

There are lessons to be learned from this episode. One is that scientists are wise to be cautious about anomalous effects. Five years after Deryagin agreed that there was no such substance as polymerized water, the journal *Science* carried a report of twin brothers in Kazakhstan who claimed to have identified a form of water that could increase the yields of cotton seeds, cause cattle to gain weight, and increase the strength of concrete. The second lesson is that, contrary to what many argued, polywater did not show that science had feet of clay, but that the 'policing' methods of science actually work. As Gould put it (1981b), 'Science is supposed to be self-correcting, and it was ... Science isn't supposed to be right all the time; it is only supposed to be able to test its claims.'

N-rays were 'discovered' in 1903 by a prominent French physicist, René Blondlot, a professor at the University of Nancy, who like Deryagin had an excellent reputation as an experimenter.[20] X-rays had been discovered eight years earlier by Röentgen, and the radioactive alpha, beta, and gamma rays had subsequently been identified also; so it might be said that the psychological atmosphere was ripe for the discovery of others. Blondlot's 'discovery' was made serendipitously while working on the properties of X-rays. In the course of these experiments he uncovered what he took to be an anomaly, and put this down, erroneously, to the presence of a hitherto unknown form of electromagnetic radiation, which he soon named N-rays in honour of his university. Blondlot went on to explore the properties of his new-found rays and began to publish his findings. This led to an outburst of research activity by others that uncovered a series of remarkable properties of the new rays. These included the 'discovery' that there were many other sources of N-rays besides the electric-discharge (Crookes) tube – for example, domestic gaslighting mantles and heated pieces of sheet-iron – and that N-rays existed in the world outside the laboratory – for example, that the human body emitted N-rays, even after death. In less than two years more than fifty papers were published on the subject. The curious fact was that all these discoveries were made by Frenchmen. Scientists in other countries were completely unable to detect the rays. There was a strong feeling among them that the N-ray was the result of honest delusion; though there were some who suspected fraud. The matter was resolved when a professor of physics at Johns Hopkins University, Robert Wood, was persuaded to visit Nancy to see for himself. Wood was a skilled hoaxer and a 'relentless pursuer of frauds such as spiritualistic mediums' (Klotz 1980: 170). Wood's report on his findings was published in *Nature* in September 1904 (vol. 70, p. 530). By close observation of Blondlot's demonstrations and some simple subterfuges, Wood was able to show that the effects Blondlot and others had claimed were indeed a delusion.

National honour was by now at stake, as were the prestige and the future careers of many of the principals. Blondlot continued to fight on. In 1905 he published elaborate instructions on how best to observe the manifestations of N-rays; it would require special training. Now, it was not the validity of the phenomena that was at stake, but the sensitivity of the observer; an argument which, as Klotz does not fail to point out, 'will not be unfamiliar to those who have followed more recent controversies concerning extrasensory perception' (1980: 175). Some defenders of N-rays even proposed that only Latin races possessed the necessary sensitivity to detect them. When, eventually, a group of French scientists proposed a definitive test, Blondlot responded defiantly: 'Permit me to decline totally your proposition to cooperate in this simplistic experiment; the phenomena are much too delicate for that.'

I have summarized this case at some length because of its relevance to our appreciation of the claims of parapsychology. One of Blondlot's later, desperate claims was that N-rays did not obey the normal laws of physics. This claim has traditionally been made by proponents of ESP – at least until the entry of the paraphysicists. It is surprising that those who object to the designation of parapsychology as a pseudo-science should pay so little attention to the manner in which possible pseudo-effects are treated in science.

CONCLUSION

We are not here to inquire what we would prefer, but what is true. The progress of science from the beginning has been a conflict with old prejudices.
T.H. Huxley

The question whether it is possible to offer acceptable grounds for distinguishing pseudo-scientific ideas from scientific ideas has been much debated. Michael Polanyi argued that there are such grounds but that it is impossible to say what they are. The 'tacit dimension' of science predominates in much of his work. I quote from a paper in which he discussed the Velikovsky case: 'A vital judgement practised in science is the assessment of plausibility. Only plausible ideas are taken up, discussed and tested by scientists. Such a decision may later be proved right, but at the time that it is made, the assessment of plausibility is based on a broad exercise of intuition guided by many subtle indications, and *thus it is altogether undemonstrable. It is tacit*' (1969: 76; Polanyi's emphasis). This comes close to saying that ideas are scientific whenever scientists say they are. Polanyi denied this (p. 75), but the denial seems to me indefensible.

Karl Popper's suggested criterion of demarcation, refutability, has often been misunderstood. It is a purely logical distinction. Whether a knowledge claim is capable of being falsified or refuted is uniquely decidable by inspecting its logical structure. One must ask: 'Is it in fact capable, in principle, of being refuted?' Whether the discovery of some fact is to be taken as actually falsifying a refutable proposition is an empirical matter. This distinction has still not been grasped by many philosophers and sociologists of science. A good example of the confusion that arises if this distinction is not understood is the often repeated contention that 'yesterday's science is today's pseudo-science'; which is a deliberately polemical way of saying that scientific ideas that were eventually rejected as false must have been pseudo-scientific – Newton's theory of optics, for example – in the same way that phrenology was eventually held to be pseudo-scientific. A related confusion is that between falsifiability and testing. A theory may be logically falsifiable but may also, at the time, be untestable because no techniques exist, or can be thought of, for testing it. The Hoyle-Bondi-Gold cosmological theory positing the universe as existing in a steady state was, for a while, in this category. It was thought to be untestable because it demanded the continuous creation of matter, but matter created in such small quantities in any given region of space as to be empirically undetectable. But other ways of testing the theory were quickly found and the theory was rejected – eventually, even, by its authors. But even if they had not been found, this would not have made the steady-state theory unscientific. The principal reason for advancing the refutability criterion is the perceived necessity to exclude non-refutable statements if scientific knowledge is to advance. It has nothing whatever to do with the question of how scientists decide what is false.

Professor Thomas Kuhn is among those who have rejected the refutability criterion, at least in respect to astrology (1970: 8). Astrology, he has argued, cannot be distinguished from science on this ground because astrological predictions can be, and have been, falsified. 'Not even astrology's most convinced and vehement exponents doubted the recurrence of such failures.' Yet, says Kuhn, astrology was not and, in so far as it is practised today, for example in India, is not a science because 'it solves no puzzles.' This remark relates to Professor Kuhn's characterization of scientific practice as puzzle solving within a generally accepted paradigm or research tradition. But this is not satisfactory. The reason astrology should be excluded from science is that it is a closed system of thought that rests on the non-refutable theory that celestial events govern the lives of individuals and even affect the course of human destiny. It is on a par with the belief of the Azande in their poison oracles. The predictions of the oracles were often incorrect, and were accepted by the Azande people as incorrect; but the oracles were part of the Zande belief in witchcraft, that is, in the proposition that witches exist – which is non-refutable.

An entirely different line of attack comes from many contemporary sociologists of scientific knowledge who (as we have seen in chapter 1) argue that all knowledge claims of whatever kind are 'socially constructed,' and that their acceptance or rejection is purely a matter of negotiation and consensus among communities of knowledge producers. The implication of this is that there can be no 'externally rational' criteria by which pseudo-scientific ideas can be distinguished from scientific ideas. The rejection, for example, of parapsychology by scientists simply means that parapsychologists have been unable to negotiate the acceptance of their ideas. But this is an entirely empty claim; it tells us nothing. One such sociologist, Dr Dolby, has, however, attempted to explicate a distinguishing principle in reference to the Velikovsky affair (Dolby 1975), which he regards as no more than a controversy over the distinction between 'serious' and 'cranky' science. This principle is the Kuhnian one, that 'heretics' must seek to establish a counter-paradigm to the prevailing paradigm and hope that it will be eventually accepted. In this case the prevailing paradigm was 'orthodox' science, represented by astronomy, geology, astrophysics, and a host of other scientific disciplines, and the counter-paradigm would be established by a community of 'Velikovskian scholars.' 'We should treat Velikovsky's ideas,' Dr Dolby says, 'as an embryonic paradigm, a coherent world view incommensurable with that of the physical sciences,' which, if it were successful in attracting talented and creative members who could 'work out the details of the new scientific world picture,' might eventually come 'to be taken seriously by the established sciences.' But this is surely taking charitableness too far, and completely distorts Professor Kuhn's intentions.

While it may be true that there can be no hard and fast line of demarcation between science and pseudo-science, there are certain features that characterize pseudo-sciences that mark them off from science. One of these is the lack of an independently testable framework of theory that is capable of supporting, connecting, and hence explaining their claims. Such a framework was altogether absent in Velikovsky's case. In the case of parapsychology, the basic assumption is that psychical phenomena exist, and there is no way in which this could be falsified: it is an irrefutable doctrine. Thus, the possibility that it may be true can never be denied, even if no convincing evidence has ever been produced to support this possibility. Even were we to assume, as many do, that parapsychologists have never produced any evidence for the existence of psychical phenomena that was not fraudulently obtained or due to statistical anomalies or the result of self-deception, we could not say that such phenomena might not properly be detected some day. This, of course, is what gives parapsychology its peculiar strength with large sections of the general public. Many people believe that because an idea cannot be refuted they have a warrant for thinking it true. But this is not so.

A related feature of pseudo-scientific ideas is that they exhibit no progress. The practice of parapsychology has been entirely devoted to repeated attempts to confirm its central dogma: that phenomena such as ESP and telekinesis exist. As to Velikovsky, Newton's laws were an improvement on what went before, if for no other reason than that they successfully generalized and explained the theories of Kepler and Galileo and went beyond that; and, in turn, Einstein's relativity theory improved upon Newton. Indeed, Einstein explains Newton, for Newton's laws are derivable from the theory of general relativity as a limiting case. The existing laws of physics, however, are not derivable from Velikovsky's ideas, which is one reason why scientists were right to reject him. If Velikovskian supporters (who include people with scientific training) wish to replace the existing laws of physics, they will have to demonstrate not only that the new laws explain everything that Velikovsky wished to explain, but also everything that the existing laws of physics are able to explain. There is no evidence that they are likely to be able to do this.

A third feature of pseudo-scientific ideas, in my view by far the most important, is that they are usually constructed in such a way as to resist any possible counter-evidence. Again, the example of parapsychology is instructive.

First, a standard argument of parapsychologists like Rhine and Soal and their associates over the years has been that the attitude of the experimenter strongly influences the psychic subject, even when the former is a firm believer in the existence of psychic phenomena. Thus, when the subject has a bad day, that is, for example, consistently fails to guess the order in which sequences of ESP cards are dealt in a manner statistically better than chance, there is a handy excuse. Experimenter attitude is said to be especially important when he or she is not a believer, and observers who are brought in to test the alleged effects (as Wood was in the case of N-rays) invariably produce a diminishing of the psychic's powers. Secondly, the acknowledged fact that psychics almost always show a marked decline in psychic ability over time (suffer 'burnout') was explained by Rhine as evidence of the 'elusiveness' of psi (his term for psychical phenomena). Thirdly, the phenomenon of a subject 'scoring a hit' (guessing correctly) on either the card ahead or the card behind, but not hitting the target, that is, not correctly guessing the card actually dealt, is described as 'forward or backward displacement,' as if this designation provides an explanation of something. In other words, the fact that the psychic subject was simply wrong is explained away. A splendid example of the tendency to exonerate the subject is to be found in a reference by Michael Polanyi to the psychic Whateley Carington. Carington is explaining why clairvoyants cannot draw a hidden picture accurately: 'Everything seems to happen much more as if those who had scored hits had been told, "Draw a Hand," for example, rather than "Copy this drawing of a Hand." It is, as one might say, the idea or content or meaning of the original that gets over, not the form' (Polanyi 1946: 22). (By 'told' he means, of course, told by thought transference from the experimenter to the psychic subject.) But if this explanation were to be accepted, then any sketch that however vaguely resembled the object in the original could be taken as evidence of the subject's psychic powers. Fourthly, it is commonly argued by parapsychologists (for example by Rhine) that psychic powers are negatively influenced by the complexity of the experimental situation. 'Experimenters who have worked long in this field have observed that the scoring rate is hampered as the experiment is made complicated ... Precautionary measures are usually distracting.' As Gardner remarks here (1980: 83), this seems to mean that the best results are obtained only as long as testing is informal and not rigorously controlled; and, in another place (1957: 207), he remarks that evidence of ESP and psychokinesis can apparently be found only when 'the experiments are relatively careless and supervised by experimenters who are firm believers.' These are only a few examples of scores of stratagems that have been used in the literature to protect parapsychological research from criticism.[21] It may be, of course, that the reasons given by parapsychologists for the shortcomings of psychic research – the peculiar sensitivity of psychic influences and their elusiveness to detection – are valid, but that is not the point. It is impossible to know whether they are valid or not.

Supporters of the claims of parapsychology, and sociologists of science who are disposed to deny the possibility of distinguishing science from pseudo-science, point to the existence of experimenter effects in science itself; to the existence of barely perceptible effects (such as gravity waves), and to the fact that scientists, historically and currently, accept anomalies and resort to ad hoc adjustments to save their theories. The difference, however, is that, in science, experimenter effects are critically watched and, to the extent possible, efforts are made to eliminate them; barely perceptible effects are regarded with suspicion and at best accepted only tentatively; and, although anomalies may sometimes be set aside as for the moment unresolvable, in the end, if they have not been resolved and the theory has not been superseded for other reasons, it will be accepted that the theory cannot be saved, since to continue to save it would constitute a block to further progress. This is not the case with pseudo-scientific doctrines; there, if the anomalies will not go away, they must in some manner be cordoned off. There are, moreover, two kinds of ad hoc hypothesizing that can be used to save theories. There are those that strengthen the theory by adding to its explanatory and predictive powers; and there are those that weaken or add nothing to the theory. The former characterize science, the latter pseudo-science.

The fourth characteristic feature of pseudo-science is total resistance to criticism. This is illustrated by Velikovsky's angry retort in an article published in *The Humanist* (Nov.–Dec. 1977) that 'nobody can change a single sentence in my books.' Among the many evasive stratagems employed are: impugning the motives of the critics; arguing that the critics are defending the status quo or 'the establishment' of science; and claiming that they, the pseudo-scientists, are the victims of a conspiracy. The general public, not being in a position to assess these claims, but on the whole taking the side of the underdog, will usually be disposed to accept them as valid. All these devices were used by Velikovsky, and he also attempted to give the arguments of the critics a psychoanalytic interpretation.

Some contemporary philosophers and sociologists of science have argued that scientists are not being true to their creed when they attack pseudo-science since science should be an 'open' culture. For example, the philosopher of science Paul Feyerabend has argued that 'there is no idea, however ancient and absurd, that is not capable of improving our knowledge' (1978: 47). If this is interpreted simply as a rhetorical plea for an open mind, it is unexceptionable. Interpreted literally, however, it is itself absurd. It is not the case that science cannot rule out in advance the possibility of any theory or any fact. It must rule out the possibility of some theories and some facts. As Morris Cohen once said, a world where no possibility is excluded would be a world of chaos (1964: 158–9).

6

Anti-science

Now, the history of the higher cultures shows that science is a transitory spectacle belonging only to the autumn and winter of their life-courses, and that a few centuries suffice for the complete exhaustion of its possibilities ... It is possible to calculate in advance the end of Western natural science ... In this very century, I prophesy ... resignation will overcome the Will to Victory of science.
Oswald Spengler, *The Decline of the West*, 1918

The gretesste clerkes ben noght the wyssest men.
Chaucer, *The Reve's Tale*

In a recent work called *The Dictionary of Modern Thought* there occur, among others of a similar nature, the terms 'anti-art,' 'anti-ballistic,' 'anti-imperialism,' and 'anti-matter,' but there is no entry for 'anti-science.' This is strange, because anti-science is much older than anti-art, almost as widespread today as anti-imperialism, and only marginally less destructive than anti-ballistic and anti-matter. Anti-science sentiments – feelings ranging from mild dislike to repugnance, fear, and even hatred – are as old as science itself. The traditional foes of science were, of course, the Christian churches, on the ground that the findings of science were generally contrary to Holy Scripture, and that the scientific attitude challenged the authority of religious dogma and undermined religious faith; but opposition from that quarter is now almost entirely confined to evangelistic fundamentalist sects. Those who have put forward philosophies based on the elevation of pure experience or intuition as the true grounds for knowledge, or the discovery of the essential nature of things through the exercise of pure reason, are generally hostile to science. Though rooted in antiquity, these ideas are far from dead. Other philosophies, such as modern analytic and linguistic philosophy, though not overtly hostile to science, have been indifferent – sometimes almost contemptuously indifferent – to it.[1] Literary and artistic 'humanists'[2] – those who speak for the arts and letters – are traditionally found among the most outspoken critics of science, the heyday of 'literary' opposition to science being the nineteenth century, when the modern manifestations of science were first becoming apparent. Social scientists – specifically sociologists, and especially those who study the sociology of knowledge – have now entered the lists. They seek to 'demystify science,' to strip away its claims to objectivity, and to reduce scientific knowledge to a system of beliefs governing the behaviour of specialized groups of people pursuing a craft.

Anti-scientific attitudes, or those traditionally possessed by the intelligentsia at least, are traceable to a variety of sources. The first and most primitive is represented by the myth of Pandora's box – the myth that curiosity can release great evils. 'In much wisdom is much grief,' said the preacher, 'and he that increaseth knowledge increaseth sorrow.'[3] A second source is the notion that science, by dissecting the world, destroys its beauty. This was the position taken by the Romantic poets, of whom Wordsworth and Goethe are representative. Goethe wrote that Newton's analysis of the rainbow 'crippled Nature's heart,' and Wordsworth that 'our meddling intellect misshapes the beauteous forms of things.'[4] The notion that science's narrowness of vision corrupts true culture was a dominant theme in nineteenth-century thought. A third source of anti-science sentiment, loosely related to the second, is the idea that the endless progression of science robs human life of meaning while at the same time it is powerless to tell us how we should live. Max Weber and Leo Tolstoy were among those who propagated this idea. But Weber had another fear: that the characteristic methods of science were symptomatic of, and perhaps largely responsible for, the growing rationality of all social institutions, which would ultimately build a cage for the future in which humanity would live in a state of 'mechanical petrification.' This idea has been very influential, especially with radical critics of technological society. Indeed, it is the identification of the pursuit of scientific knowledge with its practical application as technology (which we discussed in chapter 2), and especially its identification with the malign aspects of technology, that constitutes today the most important of all the sources of anti-science.

In part, the concern about science involves a rejection as inadequate of much in what the critics see as the logic of scientific inquiry: its stress on the regularity of *processes* over the uniqueness of *things*, its apparent refusal to admit speculative thought that is divorced from empirical test (or empirical testability in principle), its limitation of discourse to the directly communicable, its rejection of design and purpose in nature. But more than this is involved. The biologist Jacques Monod suggested that since many of the findings of modern science defy immediate understanding and intuitive, common-sense grasp – are 'beyond the understanding of most men' – they represent 'a cause of permanent humiliation.' Or, as Nietzsche said, 'All sciences today work for the destruction of man's ancient self-respect.' The poet W.H. Auden put the point in homelier terms, but cut closer to the bone perhaps, when he said: 'When I find myself in the company of scientists I feel like a shabby curate who has strayed by mistake into a drawing-room full of dukes.'

The disquiet caused by such supposed implications of science for human life has been reinforced by misunderstandings embodied in that peculiar complex of false ideas that is usually labelled *scientism* – a complex that is fed sometimes, alas, by scientists themselves. One such scientistic idea is that science produces a superior form of knowledge, a claim that derives from nineteenth-century positivism. It is invalid. There are many ways of knowing the world and the scientific way is both powerful and intellectually fruitful; but it is impossible to adjudge it 'superior' in any meaningful sense. The artist's vision and the poet's vision are no less important, and I do not believe that there are many scientists of repute today who would assert otherwise.[5] They would merely assert that science is obviously superior in that region of experience to which it relates. In any event, although science contributes to knowledge, the notion that the only true or important knowledge is scientific knowledge is clearly false.

A related scientistic misapprehension that has been very powerful has to do with the supposed infallibility of science (discussed in chapter 1): the notion that 'genuine' scientific statements must be unchallengeably true. This is a very old notion, and science is supposed to derive its authority from this fact. Vulgar opinion is still, in this sense, Baconian: that the scientific method is a means of arriving at certain knowledge about the world. But no scientist of any consequence now believes this. It is logically false, and the entire history of science demonstrates that it must be empirically false also. What is significant is the tentative nature of scientific discovery. Yet this is something that people find extraordinarily difficult to accept.

Perhaps the most important scientistic perversion, however, is to be found in the instrumentalist view of science: the view that since we can never be sure that we know the world as it actually is, all we can know is how to manipulate the world – how to predict and calculate it correctly. This view has proved attractive to the opponents of science for centuries. But it has not been acceptable within the tradition that most scientists in practice now accept. Scientific theories should not merely be useful; they must also claim to be true. Legend has it that Galileo, forced to publicly renounce heliocentrism, muttered, 'Nevertheless the earth moves.' He was not prepared to accept the way out offered him by his inquisitor[6] – assent to the argument that the heliocentric theory was simply another way of calculating the movements of the heavenly bodies – and he recanted only under duress, a method still used with notable success by those who wish to suppress ideas contrary to dogma or believed to constitute a threat to established institutions. We have seen in chapter 4 what can happen to a science (IQ psychology) that believes that computing and measuring is all that matters; and, in chapter 2, that the identification of science with technology also rests on the instrumentalist fallacy: the fallacy that asserts that science is nothing but its application in practice.

RELIGION AND SCIENCE

> The Church herself learns by experience and reflection and she now understands better the meaning that must be given to freedom of research ... We certainly recognise that he [Galileo] suffered from departments of the Church. I would like to say that the Church's experience during the Galileo affair and after it has led to a ... more accurate grasp of the authority proper to her.
>
> Pope John Paul II, 9 May 1983, addressing a gathering of scientists

Nearly a century has passed since A.D. White published his comprehensive *History of the Warfare of Science with Theology in Christendom.* It is unlikely that any serious scholar today would think it necessary to produce a new edition of this work carrying the story into the last half of the twentieth century, for science and the major Christian religions have long since reached an accommodation. All that would be required would be an appendix describing the strange phenomenon of 'creationism' (as its adherents call it), a political movement, mainly confined to the United States, that resurrects late-nineteenth-century fundamentalist opposition to the teaching of evolutionary theory. This movement has been extremely influential in recent years in local and state politics in the United States and attempts have been made to export it to Canada[8] and Britain and, through a British organization, to other parts of the English-speaking world. It is an aberration, though temporarily important, and it is opposed by leaders of the Roman Catholic church and all the major Protestant churches.

Today, anti-science is almost completely secular in spirit. Theologians, where they do not actually declare that God is dead,[9] seek to reconcile and reinterpret science and traditional religious faith in a peculiar amalgam of their own devising (the work of the late Father Teilhard de Chardin is an example). As for the scientists, they have long since learned to keep their vocation and their religious beliefs (if they have any, which many do) in separate compartments. In this way, they follow their illustrious predecessor Michael Faraday (1791–1867), who wrote: 'In my intercourse with my fellow creatures that which is religious and that which is philosophical [that is, scientific] have ever been two distinct things.'

The papacy of the sixteenth and seventeenth centuries was by no means fearful of the 'new' science; rather it embraced it as one of the ornaments of the age. The opposition came from less exalted men: from the rank and file, from the self-appointed guardians of the Holy Office, and from men who feared the social ferment attending the spread of literacy and the new knowledge that was being gained through the expansion of the known world by voyages of exploration. Copernicus's reluctance to publish was due less to fear of religious persecution than to his instinct that his theory was not quite right. He was a true son of the church; his famous work on the revolution of the heavenly bodies was dedicated to the pope (Paul III); and it was one of the princes of the church, Cardinal Schoenberg, who urged him to publish it. It was not the hierarchy that berated him, but Luther, who called him 'a fool who goes against Holy Writ.'[10] Indeed, with all the caution that is proverbially enjoined on those who make generalizations, it can be said that the effort to reconcile religion and science is, historically, characteristic of Catholic theologians[11] and scientists (in spite of the many blocks that have been thrown in their way; for example, the doctrine of papal infallibility – which extended to matters of science – proclaimed by the First Vatican Council in 1870), whereas the overt opposition to science has come most often from Protestants.

The greatest confrontations between science and religious belief came in the nineteenth century. For Newton and the scientific minds of the seventeenth century the 'works of nature' revealed a harmony and an order in the universe in which could be discerned the beneficent hand of the Creator. This world-view lingered into the nineteenth century, exemplified by such works as the Reverend William Paley's *Evidences of Christianity* (1794) and especially his *Natural Theology* (1802), both of which enjoyed great popular success. But it had already lost some of its persuasiveness in the eighteenth century, owing in part to attacks on the foundations of religious belief by the men of the Enlightenment and the scientific determinism of the French astronomer and mathematician Laplace. The most devastating blows to religious orthodoxy, and in particular to the literal interpretation of the Bible, came, however, with advances in the science of geology and with Darwin's theory of evolution. The work of erosion cannot be wholly ascribed to science, however, since the critical scrutiny of sacred texts by German biblical scholars and philologists was begun well before the impact of these scientific discoveries was fully felt. The controversy over evolution was long and bitter,[12] but it was not simply a matter of churchmen versus scientists. Scientists[13] disagreed among themselves about the theory – some on methodological grounds[14] that were quite divorced from religious prompting, but others on grounds that were avowedly religious, or religious in part. Of the latter, two of the most notable were Darwin's fellow countryman Adam Sedgwick (1785–1873), ordained minister and geologist, and the Swiss-born, but American by adoption, Louis Agassiz. 'Poor dear old Sedgwick,' as Darwin called him (though he was himself fifty years old at the time), 'looked on science as a powerful force to aid in the task of keeping mankind faithful to the word of God' (Hull 1973: 167). He was among the last of the influential natural theologians, believing that the role of science was to 'teach us to see the finger of God in all things animate and inanimate.' Agassiz was a more substantial figure as a scientist, and was the last eminent biologist to refuse to accept that Darwin had given the world of science an important theory. He remained unutterably opposed to it to the end of his life (to his great distress his son was converted), remaining convinced that natural selection was a fallacy and that species were fixed and final because they were thoughts in God's mind.

The outcome of the strife between science and religion in the nineteenth century was, essentially, to free religion from its dependence on science, and to relieve science of any obligation to values that were increasingly seen by scientists to be external to its central purpose. (It is tempting to see in this conflict the origin of the belief that science is 'neutral' and 'value-free'). Some would take this outcome to be nothing more than a kind of political concordat between science and religion in which religion got the worst of the bargain; but something more fundamental than that is involved. Science is, in its nature, agnostic rather than atheistic, though in a rather special sense that reflects a difference between the rational demonstrability of religious dogma and adherence to religious faith.[15] All religious belief is metaphysical in the sense that it can be neither directly confirmed nor falsified. But many metaphysical beliefs can be rationally criticized and discussed. The growth of science has eroded the bases of religious dogma by showing them to be very unlikely; but the notion that scientific knowledge will ever, in the words of Francis Bacon, 'suffice to convince atheism' is, in my view, simply a nineteenth-century positivist prejudice. In other words, though science may erode religious dogma (especially the dogma that the Scriptures are to be read literally because they are the word of God), it by no means follows that science can destroy religious faith. It is possible that the scientists' picture of a lawlike universe is a myth, a reflection of the structure of the human mind (although, if this were indeed so, it would still need to be explained), but it is highly unlikely. But even if it is not a myth there is no inconsistency (though of course no necessity either) in believing that the workings of the natural world reveal the hand of God – which indeed is what many scientists do still believe. It is in this sense that I say that science is itself agnostic; its findings cannot demonstrate that religious faith is warranted, but neither can they 'convince us to atheism.' The old rationalist belief that they could is simply mistaken.[16]

In his book on science in American society Professor Daniels writes that 'the coming of evolution ... finally demonstrated that no amount of "reinterpretation" could any longer hide the fact that the Bible could not be read as a scientific document' (1971: 273). This, though undoubtedly true, is precisely what the modern 'creationists' deny. They share with their fundamentalist predecessors a hostility to evolutionary theory and a belief in the literal truth of the Bible; but, unlike their predecessors, they claim that the biblical account of creation is itself a scientific document. They do not believe this, but they claim it for reasons of expediency. Their political aim has been, and continues to be, to have the biblical account of creation taught in the schools of the United States alongside the teaching of evolution as an alternative scientific theory.[17]

American fundamentalism developed as a political movement during the last quarter of the nineteenth century, though its evangelical and liturgical roots lay in the Methodist revivalism of an earlier age. It was a response to social changes that were seen as threatening the stability of what had been a basically simple, rural, and Protestant society. Immigration (bringing large numbers of people of the Catholic and Jewish faiths and of diverse ethnic origins), industrialization and urbanization, the growth of labour unions, all contributed to the perception of a creeping destruction of traditional values. Deeply conservative in the literal as well as the political sense and basically anti-intellectual, fundamentalism tended (as it still tends) to condemn all new and unsettling ideas as 'liberal' and (after 1917) 'communistic.' Its especial targets by the end of the nineteenth century were the spreading secularization of life and, specifically, the theory of human evolution that was believed to contribute to it. By the end of the century, annual 'Bible Conferences' had become forums for attacking these trends, and a number of ministers and university professors were successfully dismissed from their posts for corrupting the minds of others. For example, a geologist at Vanderbilt University was dismissed for telling his students that they were descended from organisms that pre-dated Adam. A series of pamphlets, declaring inter alia that the Bible was literally true, was widely distributed in 1910 by the Los Angeles Bible Institute, and furnished the doctrinal basis for the growing fundamentalist movement. In the first two decades of the twentieth century bills banning the teaching of evolution in public schools (the so-called Monkey Laws) were introduced in thirty-seven state legislatures, although only a handful of them became law.

One of these, in Tennessee, was challenged in 1925 in a legal proceeding that has gone down to history as the 'Scopes Trial.' John Scopes was a young high-school teacher in Dayton, Tennessee, who had been teaching evolutionary theory in defiance of the state law. When the American Civil Liberties Union decided to test the law's constitutionality, Scopes agreed to be the sacrificial lamb. The case attracted enormous national public attention, partly as a result of a unique open-air debate in front of the court-house between the two principal lawyers, Clarence Darrow for the defence and William Jennings Bryan, a fundamentalist and populist leader (and the unsuccessful Democratic candidate for president in 1896), for the prosecution.[18] Scopes lost his case (he was fined one hundred dollars), but the outcome was regarded as a defeat for the fundamentalists. Nevertheless, with the Tennessee law and two others left intact, educators throughout the United States were wary. Even as late as 1942, a nation-wide survey showed, less than half the nation's biology teachers were teaching anything about evolution to their classes (Nelkin 1982: 33). The Tennessee 'Monkey Law' was not repealed until 1967, and only then in face of defiance from many of the state's citizens. Opponents of the repeal described evolutionary theory as 'satanic' and 'likely to lead to communism.' A similar act, in Arkansas, was challenged in 1965 by another high-school teacher, Susanne Epperson, with the support of the ACLU and various teachers' associations. Mrs Epperson prevailed in the local chancery court in 1965, lost on a subsequent appeal to the state supreme court, and finally won her case in 1968 in the U.S. Supreme Court, a majority of the justices finding that the law sought to establish a religious doctrine in violation of the Constitution of the United States.

Creationism is, in part, a result of the *Epperson* judgment. Although, as Nelkin has shown, it accepts the basic tenets of fundamentalist belief, its social base is very different from that of the 'old time religion' of the Deep South. Creationists are drawn extensively from the ranks of middle-class professional or white-collar workers, and many of them work in high-technology industries. The leadership of the movement is concentrated in three principal centres: the Creation Research Society, founded in Ann Arbor, Michigan, in 1963; the Creation Science Research Center, established in San Diego, California, in 1970; and the Institute for Creation Research (1972), an appendage of the Christian Heritage College (a bible school), which is located near San Diego. These are centres for propaganda rather than, as their names would suggest, research institutions; but the names clearly reflect the creationists' desire to pass themselves off as up-to-date and 'scientific.' There are some other organizations of lesser importance elsewhere, such as the Genesis School of Graduate Studies in Gainesville, Florida (which offers a PhD in 'science creation research'), and the movement also gets support from an effective body of self-appointed textbook watchers called Citizens for Fairness in Education, whose director is an inveterate drafter of model creationist statutes (see below). The desire to associate creationism with science in the public mind is also reflected in a condition attached to membership in the Creation Research Society. This requires the possession of a master's or doctoral degree in a natural science or some branch of engineering. The mere possession of such a degree does not, of course, mean that the holder is necessarily a research scientist, but (as creationist leaders know full well) the general public are unlikely to understand the significance of this distinction. One of the founders of the Creation Research Society, and its first president, had a PhD in genetics, but he was a rosebreeder. His successor was a hydraulics engineer.[19]

A second factor attending the birth of modern creationism was the setting up in 1950 of the National Science Foundation, funded by the federal government, and the foundation's subsequent sponsorship of science curriculum reform in the nation's schools. The NSF's budget for curriculum development rose sharply from its first year in 1954 (under $2000) to a half-million dollars in 1957, and to nearly $5.5 million by 1959. The rapid growth of this program, indeed the mere entry of a federal government agency into an area previously occupied only by the states, greatly alarmed many Americans. In particular, fundamentalists were stirred by the publication in 1960 of three versions of a high-school biology textbook. These were prepared by a group of scientists from across the country (the so-called Biological Sciences Curriculum Study), and a crucial clause in the contract with the publishers gave the BSCS complete control over content (Moore 1976: 192). Evolutionary theory and material about sexual reproduction (also anathema to fundamentalists since they believe it leads to promiscuity) were included. One or other version of the book was soon adopted by most school-boards. In California, meanwhile, a document prepared by another group of scientists, *A Science Framework for California Public Schools*, became the focus of public opposition when it was presented to the state board of education in 1969. It provided the creationists with the opportunity they were seeking, and several members of the board itself argued that it should not be accepted unless creationism were given an equal place in it.

Although their ideas – and even more their tactics – are a matter of deep concern to many groups and associations, the creationists can count on a huge amount of tacit support (or at least silent acquiescence) from the public at large, for, as has often been said, 'the Americans are a religious people.' A nation-wide poll conducted by the Gallup organization in 1981 found that 65 per cent of the respondents thought that religion could answer 'all or most of today's problems'; and in another poll taken in 1982 Gallup found (to the alarm of prominent religious leaders as well as many scientists) that 44 per cent of all respondents agreed with the statement that 'God created man pretty much in his present form at one time in the last 10,000 years' (the claim made by creationists). One-quarter of these were college graduates. Of the respondents, 38 per cent said they agreed that man had evolved from less advanced forms over millions of years, but that God had guided the process. Only 9 per cent said that God had no part in the process. Catholics were more likely than Protestants to believe in evolution guided by God; Protestants were more likely than Catholics to accept the biblical account of creation. A bishop of the United Methodist Church interviewed by the *New York Times* said that the 44 per cent finding was 'almost incredible'; and the Episcopal bishop of Newark said that he 'did not know of a single reputable biblical scholar who would say that God created man in the last 10,000 years.' He called the poll's findings a 'sorry reflection on academic achievement.' The Methodist bishop put the blame on religious organizations that 'had done a poor job in teaching the meaning of Scripture.'[20]

The creationist movement is closely allied in spirit with the Moral Majority and it received the blessing of Ronald Reagan during the 1980 campaign for the presidency. At a press conference in August of that year before he made a speech to a politically oriented fundamentalist organization in Dallas, the president-to-be said that evolution was 'a theory only' and that he had 'a great many questions about it,' adding that he thought the biblical account of creation should also be taught in the schools. President Carter (whose sister Ruth Carter Stapleton is a faith-healing evangelist[21] and apparently played some part in his own 'rebirth') was asked later for his views on the matter. He said that although his own personal faith led him to believe that God was in control of the evolutionary process, 'state and local school boards should exercise [their responsibilities] in a manner consistent with the constitutional mandate of separation of church and state.'

The creationist movement has pursued four major strategic aims, the first three of which reflect very astutely certain realities of American political life: a written constitution that has been held to forbid the establishment *by any government, federal or state*, of a state church, or the passing of any law that aids one religion or all religions, or prefers one religion over another; a historically strong 'grass-roots' resistance to the centralization of state and local functions, especially education; and a long tradition of popular suspicion of intellectuals and experts, especially non-elected (and therefore non-recallable) experts. The four strategic aims are, *first*, to try to devise counter-evolutionary legislation that will pass judicial scrutiny (ultimately by the U.S. Supreme Court); *secondly*, to get and keep evolutionary ideas out of school biology textbooks or to have them rendered innocuous; *thirdly*, to persuade, and if possible to take over, elected local school-boards so that they may direct schools to give at least equal treatment to creationism; and *lastly*, to attack evolutionary theory by producing scientific evidence that throws doubt on its validity and, if possible, discredits it.

Where the attempt to forbid the teaching of evolution by law failed, as it did in the *Epperson* case in 1968, creationists sought 'equal time' for the teaching of Genesis. In *Segraves versus California* in 1981, the plaintiff (a director of the Creation Science Research Center) argued that the failure of state schools to give equal time to the biblical account of creation infringed the religious freedom of his child, Kasey Segraves, and was therefore unconstitutional. The court rejected his plea. When it seemed that the demand for 'equal time' would not pass muster in the federal courts, the creationist leaders invented the notion of 'creation-science' and claimed that it should receive 'balanced treatment' alongside of the teaching of 'evolution-science.' This claim, that creationism has the status of a science, was subjected to close judicial scrutiny in the case of *Reverend McLean versus Arkansas* in 1981. It was the most important trial for the anti-evolutionists since *Epperson*, partly because Arkansas is one of the traditional strongholds of anti-evolutionary public sentiment. At stake was Act 590 of the Arkansas legislature, signed into law by the state governor in March 1981, 'An Act to require balanced treatment of Creation-Science and Evolution-Science in Public Schools.'[22] The governor, Frank White, a member of a small evangelical bible sect who had been supported in his campaign for governor by the Moral Majority, publicly admitted that he had not read the act before signing it. Many of the state legislators who voted for it were dismayed when it passed: 'This is a terrible bill, but it's worded so cleverly that none of us can vote against it if we want to come back up here,' said Representative Bill Clark.[23] The bill's sponsor, a state senator, was a 'born again' Christian fundamentalist. There was no consultation about the bill with the state department of education, science educators, or the state attorney general. It was passed after only a few minutes' discussion in the senate, and in the house of representatives following a fifteen-minute committee hearing.

After a lengthy trial in which expert and not-so-expert witnesses were called by both sides (one of the defence witnesses, a teacher at Dallas Theological Seminary, startled the courtroom by stating that there is nothing necessarily religious about God, and amused it by admitting that he believed in the existence of UFOs because this had been 'confirmed by an article in *Reader's Digest*'), the trial judge, Judge William R. Overton, found for the plaintiffs. The statute contravened the First and Fourteenth amendments to the Constitution of the United States, he said, because it was not purely sectarian in purpose, was intended to 'advance a religion,' and would foster 'an excessive government entanglement with religion.' He added *obiter dicta*: 'The application and content of First Amendment principles are not determined by public opinion polls or by a majority vote. Whether the proponents of Act 590 constitute the majority or the minority is quite irrelevant under a constitutional system of government. No group, no matter how large or small, may use the organs of government, of which the public schools are the most conspicuous and influential, to foist its religious beliefs on others.'

The creationists were not deterred by the Overton judgment from pursuing their legislative aims; indeed, even before the court's decision was handed down, work had already begun on a revision of the model bill prepared by Paul Ellwanger, the director of Citizens for Fairness in Education (which with very minor changes became Arkansas Act 590), substituting the principle of 'unbiased presentation' of differing accounts of origins for the principle of 'balanced treatment' of creation-science and evolution-science. Ellwanger has publicly avowed that his aim is to get a form of wording that will eventually succeed in the courts; and this is believed by some creationists to be possible if they can, in the end, convince a wide enough uncommitted public that they are a persecuted group.

Since the Arkansas case the Supreme Court of the United States has struck down a 'balanced treatment' statute in Louisiana by a seven to two majority (the chief justice and one other justice dissenting). The comment of the executive director of the Institute for Creation Research was that the decision would intensify their efforts to bring creationism into the schools by persuasion (see below).

The second line of attack for creationism has been the high-school textbook, and this requires the persuading or pressuring of state boards or departments of education. Here creationists have been notably successful. For example, since 1974 the Texas State Board has required that all science textbooks seeking adoption by state schools must identify the theory of evolution as 'only one of several explanations of the origins of humankind.' All textbooks 'shall be edited, if necessary, to clarify that the treatment [of evolution] is theoretical rather than factually verifiable.' In addition, 'each textbook must carry a statement on an introductory page that any material on evolution included in the book is presented as theory rather than fact,' and 'the presentation of the theory of evolution should be done in a manner which is not detrimental to other theories of origin.' Texas is an 'adoption' state (that is, one of about twenty states of the Union with textbook commissions that determine what books are to be used in the state schools) and it has a very large school population. States like Texas are immensely important to textbook publishers because they provide a large guaranteed market. This means that publishers are sensitive even to their detailed requirements about content; it also means that creationists will tend to concentrate their attentions on the larger adoption states (as they also did in California). Nelkin quotes two publishers: 'We are very, very aware of the concern of scientific creationism, which is what the moral majority wants ... We can't publish a book in 1983 that doesn't recognise another point of view.' And, even more sadly: 'Creation has no place in biology books, but after all we are in the business of selling [them]' (Nelkin 1982: 154). There are, however, certain counter-pressures that reflect court decisions about the constitutionality of creationist statutes, since the issue of what may legally be taught in the schools obviously bears on the nature of the textbooks used in teaching. An example is a decision by the board of education of New York City following the Overton judgment, that three biology texts in current use in city schools were 'unfit for use' because of the inadequacy of their treatment of evolution. Hence, we have the spectacle of the major school textbook writers and publishers warily watching the progress of a battle on two fronts: between creationists and 'civil libertarians' and other opponents of creationism in the courts, and between creationists and anti-creationist professional educators and education administrators in the state capitals.

The third line of attack for creationists is through the elective school-boards at the local level and through the schools and teachers themselves. They have had a fair amount of success in persuading local electorates to vote for members of school-boards and elective officials who are sympathetic to the creationist cause, and in organizing local lobbies to pressure the public. They have also been prominent in an 'alternative schooling' movement that promotes opting-out of the state system through the establishment of schools that exclude science teaching altogether and 'promote moral and religious values.' There is no doubt that many teachers have been intimidated by these pressures; and there are many others who simply avoid the possibility of 'hassles' with parent-teacher associations by omitting any reference to evolution in their class-rooms, even though they know they have the backing of the National Association of Biology Teachers. In some areas bolder teachers have formed local coalitions with anti-creationist parents and clergy; for example (Nelkin 1982: 156), the deliberately, but unfortunately, named Georgia Ontological Association for the Protection of Evolution (GO APE).

In his excellent and temperate book *Abusing Science: The Case against Creationism*, the philosopher Philip Kitcher emphasizes that the creationist movement is not just an attack on evolutionary theory. Because science is a seamless web of ideas and discoveries any attack on evolutionary theory is also an attack on geology, molecular biology, large parts of physics and astronomy, palaeontology, and palaeoanthropology (1982: 4). One can go further. As part of a more general contemporary campaign against science, its 'know-nothingness' and irrationalism threatens all genuine scholarship.

Creationists use instances of scientific disagreement quite unscrupulously. Disagreement for them is somehow discreditable – or at least that is how they wish to present it to the public. They employ the tactic common to all anti-scientific and pseudo-scientific groups: that science cannot 'prove' that a theory is correct (this is true but irrelevant). 'If you can't be certain,' they say, 'then how can you make the claims that you do?'[24] We saw in chapter 5 that this argument was used by supporters of Velikovsky; we also saw why it is invalid. Creationists also imply that all evidence that is critical of Darwinian theory somehow supports their case; but this is not so, for the fact that a theory is false does not mean that its contrary is true; both may be false. In fact, the real case for creationism must rest on revelation, on the literal truth of the biblical account, and not on empirical evidence; for as creationists themselves have said, 'We do not know how the Creator created, what processes he used, for He used processes which are now not operating anywhere in the natural universe ... We cannot discover by scientific investigations anything about the creative processes used by the Creator.'[25] 'If man wishes to know anything about creation ... his sole source of true information is that of divine revelation. God was there when it happened. We were not there.'[26] These passages are alone sufficient to expose the fraudulent nature of the creationists' claim that their account is a scientific alternative to evolutionary theory; and the assertion that God's creation cannot be explained is accompanied by strenuous attempts to refute the evidence for evolution, even though creationists repeatedly state that the theory of evolution cannot be refuted and is therefore as much an act of faith as their own alternative 'theory'![27]

Their major lines of attack on the evidence for evolution are extensively documented and analysed in the literature listed in note 17, so I shall confine myself here to a very brief summary. All their arguments rest on misunderstandings of scientific principles or deliberately misrepresent them. The first is the claim that any evolutionary process in nature is precluded by the second law of thermodynamics (which can be interpreted as a law of increasing disorder or randomness), since the evolution of life requires the emergence of living forms of increasing order and complexity. The answer to this is that the second law applies only to closed systems and that the earth's biosphere is not a closed system (life depends on energy from the sun and the sun's energy comes from outside the biosphere). Some creationist apologists are aware of this and deal with it by changing the subject (Kitcher 1982: 93). Living things are a standing challenge to the second law of thermodynamics, but they do not violate the law any more than astronauts violate the law of gravitation. Life is a struggle to preserve order in face of a law of nature, and surplus energy is required to do this.

The second argument is that the emergence of life from non-life is so improbable as to be impossible, given the statistical laws of probability. 'The real problem for evolutionists is explaining how a cell in all its complexity could arise suddenly from simple inorganic atoms ... The sudden "poof" formation of a cell would demand a supernatural act by an agent with supernatural power and intelligence' (Wysong 1976: 410–11). But no evolutionist claims that life arose 'suddenly.' On the contrary, the argument is that the age of the earth is long enough – some 4.5 billion years – for the gradual emergence of life from non-life to have been almost inevitable. Indeed, we may nowadays ask where the line is to be drawn between life and non-life. A virus is inert outside a cell: a mere chemical. It only replicates itself within the cell. Is a virus a living thing?

The notion of *chance* is also misused or misunderstood by creationists. It is only in an almost metaphorical sense that we speak of life as having evolved 'by chance,' for we know that changes from one state to another do not take place entirely at random, but only within a framework of possibilities set by preceding states. Creationists also point to quantum mechanics, and especially to Heisenberg's uncertainty principle. But the fact that subatomic phenomena must be interpreted probabilistically does not mean that supra-atomic phenomena must be so interpreted. On the contrary, the application of quantum theory to chemistry has allowed organic chemists to explain why the elements regularly combine in precisely the ways that they do.

A third, and connected, set of arguments relates to the age of the earth and the fossil record. The age of the earth has been subject to continual upward revision by scientists since the beginning of the nineteenth century, but the advent of radiometric dating has made its computation accurate and very reliable. Since every radioactive element in the earth's crust decays at its own characteristic and constant rate, these elements together constitute a kind of self-correcting geological clock (one rate checking another). Creationists have constantly challenged the use of carbon-14 dating since the isotope has a relatively short half-life (about 6000 years), which means it cannot be used to date beyond about 50,000 years. There are, however, many other much longer-lived radioisotopes that are used in dating. Uranium, for example, yields billions of years for the oldest known rocks. In order to combat the computed age of the earth using such methods (about 4.5 billion years), creationists have been forced to argue that before a certain time, which they usually associate with the Flood about 6000 years ago (a central event in their doctrine), these decay rates were much more rapid than they are now. Hence scientists are misled (by their adherence to constant rates of decay) into believing that the earth is much older than it really is. There is no scientific evidence for this slowing down of decay rates. It is worth noting that all the geological witnesses called by the creationists in *McLean versus Arkansas* admitted that this is the case. Moreover, as one scientist (quoted by Cracraft 1982: 82) has pointed out, if the earth were only a few thousand years old as the creationists claim, their pre-Deluge radioactive decay rates would have been so great that the heat generated would have been sufficient 'by a large margin' to have vaporized the earth.

As for the fossil record,[28] the creationists claim both that there are gaps in it (which no scientist denies) *and* that it is nevertheless sufficiently representative to throw doubt on evolution. There are good explanations for the gaps (apart from the obvious fact that only a fraction of the fossil record has yet been uncovered). And the existing record is not representative; on the contrary it is biased – and for the same general reason that explains the gaps, namely that some kinds of organisms are more likely to survive as fossils than other kinds, and some kinds – with soft bodies – will leave no trace. Much can be learned from the existing fossil record, but some parts of it are better than others. The creationists' strategy, as Kitcher says, is to ignore the good parts and focus on those groups of organisms that scientists know, on independent grounds, are unlikely to leave fossilized remains.

The care and sincerity with which creationists scrutinize and use scientific evidence is further called in question in the now famous case of the Paluxy Creek footprints. In 1939 a palaeontologist named R.T. Bird published an account of his discovery of dinosaur tracks in a creek on the Paluxy River in Texas. It has long been rumoured that additional tracks were manufactured in the 1930s by some local residents of nearby Glen Rose in order to make the place more viable as a tourist attraction. Creationists have claimed that certain tracks in the river bed alongside the (genuine) dinosaur prints were made by human feet, showing, of course, that Man and dinosaur were contemporaries. In the early 1970s a film was made about these prints by a body called Films for Christ, and the film has had a wide showing. The depressions in the rock were made to stand out better for the film-makers by touching them up, possibly with oil. This film has been closely examined with stop-frame equipment by a physical anthropologist at the University of Massachusetts, and she has shown that when the added outlines are removed from the pictures the illusion of human feet is destroyed. She subsequently made an on-site inspection and discovered that the prints show irregular stridelengths and change size and even direction with every 'step.' It is now almost certain that these strange effects were simply the result of erosion, with perhaps some deliberate alteration also. The phenomenon on which they once placed so much emphasis has become something of an embarrassment to creationists, and some have finally disowned it.

The creationists have also directed their attentions to science museums. There is no doubt that the business of selecting and displaying objects in a museum of any kind – be it a science museum or a museum of modern art – is value-loaded. In particular, a science museum normally shows only scientific successes and gives little if any attention to what is *not* known about the subject or is considered at the time to be of doubtful validity. Creationists sometimes make use of these facts. In 1978, for example, they attempted (though without success) to block an evolution exhibit at the Smithsonian Institution in Washington, DC; and, in 1982,[29] British creationists brought enough pressure on the Museum of Natural History in London to cause the museum authorities to display a placard outside its 'Origin of Species' exhibition stating that creationism held that living things had been created 'perfect and unchanging.' Thus, as one writer has sardonically observed, 'evolution was discredited at the entrance' (George 1982: 140). The journal *Nature*, reporting this amazing phenomenon, announced 'Darwin's death in South Kensington.' But Darwin (or rather, the evolutionary principle – for that is what in truth it is) will not be done to death by placards, or for that matter by censorship and legal prohibition. The *principle* of evolution is now beyond dispute. The task begun by Darwin but not completed by him is to explain its mechanisms, to understand the process better. Creationism attacks 'evolutionary theory'; but there is no settled evolutionary theory, only an evolving theory about evolution.

THE INTELLECTUAL REVOLT AGAINST SCIENCE

I come to cast off Bacon, Locke, and Newton ...
William Blake (1757–1827)

It is no longer feasible, or indeed in a democratic society defensible, for science to remain unscrutinized or for scientists to resist the pressure to hold them answerable for what they do. This is not an argument for the public control of science (whatever that might mean in practice), still less for the public's right to decide what is to count as scientific or what should be admitted as scientific knowledge. But there are areas of scientific practice today where questions (moral, 'humanistic,' and political) can legitimately be asked, and which it is, in fact, to the long-run advantage of science itself to have publicly debated (as the existence of organizations of 'concerned scientists' would seem to testify). Such critical questioning is not 'anti-scientific' and should not be treated as such. It is not, for example, anti-scientific to raise moral and legal doubts about the use of *in vitro* human fertilization techniques, or to question the wisdom of certain types of genetic engineering, or to challenge the use of live animals for experiments when other methods not involving animal suffering are available;[30] or, even, to question the funding priority given to some particular areas of research over others. It is true that the questioning of scientific practices (as well as the consequences of science in technological application) has been used by some to condemn the entire scientific enterprise, just as Ivan Illich in his book *Medical Nemesis*[31] used the genuine and rightly alarming problems of unnecessary surgery[32] and iatrogenic illness[33] to attack the whole of modern medicine – advising his readers to abandon it and to return to 'natural' methods. In that sense such questioning becomes anti-scientific. Here, however, I shall confine the use of the term (as I did in discussing creationism) to attacks on the fundamental bases of science.

The grounds for the intellectual revolt against science, from the traditional 'humanistic' to the contemporary philosophical and sociological, can be summarily reduced to three. First, that science 'holds a monopoly on the totality of all phenomena occurring in nature' (Nowotny 1979: 16). That is to say, scientists legislate what phenomena are to be considered 'natural.' This raises the demarcation issue discussed briefly in chapter 5. Hadrons[34] and gene transference are natural phenomena; extrasensory perceptions are not. Secondly, that science 'holds an increasingly powerful monopoly of access to the investigation of nature.'

That is to say, the world can be investigated only by (increasingly sophisticated) techniques and instrumentation that are inaccessible to the non-scientist. This implies that science must exclude all those who seek communion with nature through direct sensory experience, and who reject the separation that science makes between the observer and that which is observed. This is a criticism of method, whereas the first might be considered an ontological objection. Nowotny's use of the word 'monopoly' in both these instances indicates the critics' sense that science is becoming increasingly successful in excluding serious consideration of any world-view other than its own. The third ground of objection (which Nowotny does not mention) is confined to followers of the idealist tradition in philosophy from Hegel through Husserl to the Frankfurt School and Habermas. It is that modern science (science since Galileo) is defective in its lack of a critical reflexivity:[35] that it fails to reflect on its own foundations, to accept that it deals only with the surface of things and not with the underlying essential reality – and that this is a necessary consequence of its method. This argument may sound excessively metaphysical and remote (it is, of course, intended to restore the hegemony of metaphysics, which that discipline so long enjoyed), but it has some practical consequences, one of the more interesting of which is that it gives support to the ideas of contemporary sociologists (and some philosophers) of science who wish to persuade us that science is nothing but a system of beliefs, a mythology conveying a set of ideas that are internally consistent, that help to make sense of the world (at least for scientists), but that bear no more knowable relation to reality than does a mythology of witchcraft possessed by a primitive society.[36]

The aim of the critics is quite explicitly to cut science down to size – exemplified in the philosopher Paul Feyerabend's deliberately provocative question 'What's so great about science?' (1978b: 73) yet most of them seek not merely to refute the scientistic pretensions that are sometimes made on its behalf (for example, that it is the sole source of 'reliable' understanding of the world), but to demystify it and, by so doing, to change it. Feyerabend is a leading exponent of this policy. His recent work constitutes a sustained attack on the foundations of science (all three of the above-mentioned objections are present in it), making use of an elaborate set of ingenious arguments for 'epistemological anarchy' – for a scientific methodology in which 'anything goes,' in which science would be open to any and all kinds of influences and to all modes of thought and enquiry. His *Against Method* (1978a) and *Science in a Free Society* (1978b) are full of epistemological and practical claims. For example, there is no idea, however ancient and absurd, that is not capable of improving our knowledge; political interference to force science to admit alternatives is justified; the scientific world-view has triumphed because it has been given the imprimatur of the state; it is closer to myth than the philosophy of science is willing to admit; scientific rationality is only one kind, and in many respects an inferior kind of rationality; science is anti-democratic because (among other things), while citizens are free to choose their religion or to reject all religions, they are not free to accept or reject science; dissenters to 'established' science are routinely suppressed, therefore lay persons should 'supervise' science; and so forth. Feyerabend's writings have exerted a powerful influence, often on minds greatly inferior to his (for example, some of the contributors to Nowotny and Rose 1979).[37] He has actually alarmed some otherwise radical critics, for example Hilary Rose (1979), who wonders whether 'reflexivity' and the demythologizing of science have not been pushed too far.

It is one thing to criticize science, another entirely to project what a reformed science might look like. The transfiguration of science was a continuing theme in the work of the Frankfurt School and especially in the writings of Marcuse. Their claim, as we have seen, was that 'classical' science was quite different from 'modern' science in that it was metaphysical, concerned with eternal truths, accepting of and at one with nature, and inherently 'liberating,' whereas 'modern' science is anti-metaphysical, concerned only with measurement and successful prediction, wholly instrumental in its approach to nature, and directed towards domination. It was never made clear, however, whether a 'transfigured' science meant a possible change in the nature of science itself, or if it simply referred to the way existing science would be put to use in a transformed society. Contemporary counter-movements to science have done nothing to remove this ambiguity, as indeed the various terms that are used make clear: a 'new' science, a 'liberated' science, an 'alternative' science, a 'transformed' science. Different critics argue for different solutions: science substantially as it is, but undertaken in a more 'humane' spirit; a science more open to ideas and influences that it now excludes; a transformed science operating within the confines of existing society; or a transformed science that would result from a radical restructuring of social institutions.

The classic historical case of an alternative science is that of *Naturphilosophie*,[38] which flourished in Germany in the early decades of the nineteenth century. It was a true 'alternative' in that it constituted a coherent system of ideas and principles, in effect an alternative world-view, unlike mesmerism and phrenology,[39] which roughly coincided with it in time. Though primarily a German phenomenon, nature philosophy exerted some intellectual attraction for the English Romantic poets, especially Samuel Taylor Coleridge. More important, a number of prominent scientists, including Oërsted (the discoverer of electromagnetism) in physics, Oken, Richard Owen, Müller, and (in his early years) Agassiz in biology, and Humphry Davy in chemistry, came under its spell to a greater or lesser extent. It was associated originally with Schelling (1775–1854), a confidante of the young Hegel, and was expounded by him in two principal works: *Ideas toward a Philosophy of Nature* in 1797, and *First Outline of a Philosophy of Nature* in 1799. The core of Schelling's doctrine was that objective scientific knowledge is possible, but can be obtained only by means of a philosophical investigation of what can be known a priori. As one commentator puts it (Knight 1975), the nature philosophers believed that scientific discoveries could be made from an armchair, simply by taking thought. This was Hegel's view also. As is well known, Hegel 'proved' by pure reason that no more than seven planets could exist just at the time when astronomers were on the point of showing, by observation, that he was wrong. This 'quixotic attack of the young Hegel [on science] ... and his swift defeat at the hands of the scientists' has been described by Polanyi (1958: 153) as one of the great formative experiences of modern science.[40] This is not to say that the practising *scientists* who adhered to *Naturphilosophie* were opposed to experiment. They were not, but the interpretation of experiment was always done in the light of the a priori principles of *Naturphilosophie*.

The greatest philosopher of nature in the western world, Goethe (1749–1832), was not an adherent of *Naturphilosophie*, though he was to some degree influenced by it; nor was he, even, a philosopher in the professional sense. He was a man who approached nature and studied it with the eyes and mind of a poet: the archetype humanist-scientist sought by many of today's critics. He appears to us not simply as a man of letters who happened also to be a naturalist. His poetry, novels, and plays were suffused with his peculiar understanding of, and reverence for, nature; his treatises and essays on botany, geology, meteorology, physiology, and many other subjects, and the observations and experiments on which they were based, were shaped by his poetic sensibilities. His science and his letters stood in a completely symbiotic relation. It is easy to see why Goethe-as-scientist has for so long fascinated opponents of science-as-it-is. It has been well said (Gillispie 1960) that Goethe's 'nature' was not objectively analysed by him, but 'subjectively penetrated.' Nature had to be comprehended whole – or, as we would say nowadays, holistically. The central core of *Naturphilosophie* was that there is a single unifying principle underlying the whole of nature and bringing all its parts into harmony. Goethe wished to know 'what nature contains in its innermost depths.'[41] The observer is not detached from nature; he is part of it. 'That is the trouble with modern physics,' Goethe wrote, 'it wants to perceive nature only through artificial instruments. Man himself, using his healthy senses, is the greatest and most exact physical instrument which there can be.'

The analytic method of science is rejected. Chemistry 'only destroys by analysis.' Goethe arranged objects according to their aesthetic aspects, not according to Linnaean principles. Gillispie asserts (p. 139) that for Goethe 'there was in Linnaean botany nothing but nomenclature, nothing for the intelligence nor the imagination, no place for loveliness of form and flower.' Linnaeus 'turned the stream of life into a mosaic of skeletons.' Newton's analysis of light by use of the prism is the best-known example of Goethe's rejection of 'anatomising' and dissection. But his attack on Newton's work on optics embodied a misunderstanding of what Newton was about, however wrong Newton may have been about the particulate constitution of light. Goethe's own experiments with the prism produced a treatise on colours, not a treatise on light. His *Theory of Colours* was theorizing about the physiology of our perception of colours. He believed (correctly as it turned out) that 'the eye does not delude itself,' but that our interpretations of what the eye yields to us may do so. However, since his interpretations of what he saw through his prism were shaped by his preference for the aesthetic, he produced a set of ideas that were of value to artists but could not form the basis on which physical theory could advance. If Goethe's approach to science had been followed, if his methods had been consistently applied in the physical sciences and in biology – indeed in so far as, for a time, they *were* applied – they could only lead science astray. Many of today's sociologists and historians of science would no doubt argue that all this means is that a counter-movement that would have established a different kind of science was defeated by the sheer weight of 'established' science. This seems to be the implication of Gillispie's remark (p. 199) that the present 'survivals of romanticism' (that is, the counter-movements to science) are 'relics of the perpetual attempt to escape the consequences of western man's most characteristic and successful campaign [that is, science], *which must doom to conquer*' (emphasis added). But such sentiments, which are widespread, perpetuate what I believe to be a false polarity – that between a bloodless, measuring, numbering, dissecting science and a qualitative, subjective, 'humane' understanding of nature.

The Victorian physicist, and expositer of science for the laity, John Tyndall wrote: 'Presented rightly to the mind, the discoveries and generalisations of modern science constitute a poem more sublime than has ever yet addressed the human imagination. The natural philosopher of today may dwell amid conceptions which beggar those of Milton.' Thought by many of his contemporaries to be expressing the essential creed of the new 'scientific naturalism' to which they were unalterably opposed, these words seem even truer today than when they were written. To give but one example: the invention of the electron microscope vastly widened and deepened the scope of the natural world; it revealed a subterranean world that was totally unknown, and for the most part unsuspected, before. Its use uncovered images of startling beauty and wonder every bit as moving as Wordworth's golden daffodils and Goethe's rainbow.

The romantic view is not without a kernel of truth, but the claims that are made for it are sometimes ludicrous. Consider, for example, the overheated words of a historian of science, E.A. Burtt:

> It was of the greatest consequence for succeeding thought that now the great Newton's authority was squarely behind that view of the cosmos which saw in man a puny, irrelevant spectator ... of the vast ... system that constituted the world of nature. The gloriously romantic universe of Dante and Milton ... had now been swept away ... The world that people had thought themselves living in – a world rich with colour and sound, redolent with fragrance, filled with gladness, love and beauty, speaking everywhere of purposive harmony and creative ideals – was crowded now into the brains of scattered organic beings. The really important world outside was a world hard, colourless, silent, and dead. (Burtt 1924, 1932)

But what was this world of fragrance, gladness, and love, the world that 'people' before Newton inhabited, really like? The world of Dante was a world of ignorance, fear, and superstition for the poor and the majority of the rich alike. It was a world in which, shortly after Dante's death, more than half the population of Europe was decimated by bubonic plague. The 'gloriously romantic universe' of Dante and Milton was a world in which many thousands of women were tortured and done to death as witches. It was a world in which Milton lived to see his books burned for his support of the Puritan cause. It would be interesting to hear in what ways, in principle, the violence, social disturbance, and cruelty of those times of 'colour and sound' differed from our own.

One of the gurus of the media-styled 'counter-cultural revolution' of the 1960s, Theodore Roszak, is reported as having said (Chedd 1971) that 'the great cultural project of the next few generations [would be] to produce a synthesis between science and other modes of consciousness, which include mystical traditions, aesthetic experiences and many others.' In his principal writings (1968, 1972) Roszak toyed nostalgically with the Hermetic, Cabbalistic, and neo-Platonic traditions that existed alongside science as it emerged from the Renaissance, implying that science lost much that was important when it cut itself free from them. To the extent, today, that we are scientific, he said, we are not truly human. His heroes, not surprisingly then, were William Blake, Goethe, and the Romantic poets, his goal the 'visionary experience,' his teachers the shaman and 'the magic of the environment.'

A dissenting scientist, the chemist Thomas Blackburn, writing in *Science* in 1971, argued that the salient feature of the counter-culture was its 'epistemology of direct sensuous experience, subjectivity, and respect for intuition – especially intuitive knowledge based on a "naive" openness to nature and to other people.' His contention was that science must change because it had been shown to be incomplete; incomplete because it failed to accommodate to this epistemological mode (Blackburn 1971). He asserted that there were three 'tenets of counter-cultural thought that [held] great promise for the enrichment of scientific practice and, perhaps, for the improvement of scientific morality.' These were that the 'most reliable and effective knowing' follows from direct and open confrontation with phenomena; that since the self and the environment are inextricable, 'one can understand his surroundings by being sensitive to his own reactions to them'; and that 'because knowledge of nature is, in this way, equally open to all, the expert is highly suspect.' It will be seen that the ideas of Roszak and Blackburn relate closely to Nowotny's criticisms of science.

What then – realistically – are the prospects for an alternative science that would 'synthesize' science and other modes of consciousness (Roszak) or 'accommodate' it to a different epistemological mode (Blackburn)? The answer is 'none' – for such an 'alternative' would no longer be science. The practice of science at its best, as many eminent scientists have testified, demands vision, intuition, a respect for the natural world, a sense of the wholeness of things, even a sense of mystery; all things that the 'humanist' values. But the 'common-sense theory' of knowledge that is the basis of Blackburn's three tenets of counter-cultural thought – the theory that nature is an open book that can be read by all who submit to it – has long since been rejected by science, and its success has depended upon this rejection. At the root of the scientific approach to understanding lies what the biologist Jacques Monod once called 'the most powerful idea ever to have emerged in the nöosphere': the idea of *objective knowledge*. Science postulates the existence of a world about which it is possible to have *public, directly communicable* knowledge. This must be expressed in precise, unambiguous language. The ineffable, the mystical, the sensual cannot be expressed unambiguously; any attempt to do this would destroy what the speaker or writer is seeking to convey.

All social knowledge is intersubjectively transmissible in a trivial (indeed tautological) sense, but it is not the same intersubjectivity that characterizes science. The mystic may say, 'Under my guidance you may attain to a knowledge of the inner – or true – meaning of things, such as I possess'; but this knowledge is, and must remain, essentially private to the experiencing person. The scientist assumes that the world that he or she experiences is *exactly the same world* as other scientists experience. The language of the poet, or the visual 'language' of the artist and the sculptor, is expressive. It seeks to convey, and perhaps release, certain feelings and emotions and to communicate ideas through them. The language of the scientist is argumentative. Scientific discourse is argument, not conversation.[42] It is not 'edifying' discourse either.[43] Nor is it dialogue aimed at arriving at a synthesis; it is informed by a two-valued logic in which an idea must either be true or false and the affirmation of one proposition entails the denial of its contradictory.

The public character of science and the precision with which its propositions are expressed are prerequisites for open inquiry and for the criticizability (and criticism) of ideas. Yet science is also a *disciplined* activity. It is necessary for scientists to agree on many things, to take much for granted, even if only provisionally. It is necessary because science is a collective enterprise. Progress in science would be impossible if everyone 'did their own thing' (as Feyerabend seems to recommend). It depends, rather, on many people doing the same thing. There is conflict in science (especially of ideas), but there must be agreement too. For example, as John Ziman reminds us, the collective nature of science restricts scientific information about the external world to those observations on which independent observers can agree. The assumption that all observers are equivalent 'is not merely a basic principle of Einstein's theory of special relativity; it is the foundation stone of all science' (Ziman 1978a: 43).

Lastly, the much criticized 'abstractness' of science is the core of its strength. Science is speculative, conjectural, but it is nevertheless firmly rooted in an empirical base. The Greeks were more interested in abstract speculation than they were in empirical evidence. Other civilizations – early Islam for example – placed great emphasis on accurate observation and experiment but gave little attention to the development of theory. It was the major achievement of the scientific revolution begun by men like Galileo to have succeeded in bringing together abstract theorizing and empirical inquiry in a single undertaking. It is this peculiar conjunction of abstract speculation with empiricism that distinguishes science from myth and from purely empirical (or almost purely empirical) enterprises such as alchemy. Science is conjecture checked by evidence; but conjecture is paramount. A scepticism about what appears immediately to be true and an imagining of the ways the world may be: these are the hallmarks of science. Both the scientist and the 'humanist' deal in imagined worlds. The difference between them is that, as Sir Peter Medawar once said, 'the business of science consists in trying to find out if [the scientist's] imagined world is anything like the real one. If it is not, then we have to think again.'

Spengler's gloomy prophecy of the end of science, with which I began this chapter, is unlikely to be fulfilled. Some scientists have lost their faith, but they constitute a very small minority.[44] The 'will to victory' of science seems as strong as ever. Numerous opinion surveys taken over an extended period confirm that a majority of the general public remain constant in its support. Science will not be much affected by those intellectuals who attack its methodological and epistemological foundations on 'humanist' grounds. The more substantial threat to science – or perhaps one should say (stressing pure science) to the scientific spirit – comes from trends in its own organization and practice, from the greatly increased militarization of pure research and, more generally, its increased subjection to practical utility.

Notes

CHAPTER 1

1 I am referring here to the fact that in many primitive and traditional societies it has been believed that ideas, expressed in words, do have the power to alter and control physical reality, and that they even have the power, uttered under appropriate circumstances, to bring about the events or states that they denote.
2 Hegel's variant on this was that individual minds are only pieces of a single mind, the Absolute.
3 Solipsism cannot be refuted, but it is absurd. It is possible that nothing exists outside of *my* mind and that everything I think exists is an illusion. But I do not believe this. If I did, I should not be writing this. Bertrand Russell recalled (in his *Human Knowledge: Its Scope and Limits*) that he once received a letter from 'an eminent logician' saying that she was a solipsist and was surprised that there were no others like her. But perhaps she was joking.
4 Realism and Idealism are contradictory doctrines, not contrary doctrines as is sometimes said. If Realism is true, Idealism must be false, and vice versa. Two contrary doctrines may both be false.
5 A formerly extremely influential epistemological theory. The philosopher Karl Popper called this 'the bucket theory of the mind'; that is, that the mind is an empty vessel into which sense impressions flow.
6 Or, at least, groups of atoms – which appear as white patches on a photographic plate (see, for example, *New Scientist*, 28 May 1970: 414).
7 And also Popper (1962), 118–19
8 This in essence is the core of Popper's 'World 3' to which reference will be made later.

9 John Locke, *Collected Works* IV: 145
10 See especially Husserl 1970.
11 Such problems may, of course, vary from the very large and difficult to the relatively small, but all should be seen as soluble in principle, if not immediately with the available knowledge and resources. Problems that are set aside as currently insoluble are likely to be revived eventually if they are important enough. Newton recognized that universal gravitation should long since have drawn all the matter in the universe into one place. He was unable to propose a solution to this problem but it has been solved. Olbers's Paradox is another example of a cosmological problem. In brief this asks, since the universe is so immense and contains so many stars emitting light, 'why is it dark at night?' This problem remains unsolved, though many solutions have been attempted (see *Scientific American* 288, no. 4, 122A).
12 Fawcett (1833–84) was an economist, friend of Darwin, and a staunch defender of his theory of natural selection.
13 Lyell (1797–1875) was one of the most famous geologists of the nineteenth century. Initially opposed to Darwinism, he was later converted.
14 A distinction is sometimes made between hypotheses, theories, and laws, hypotheses being regarded as rather lowly theories, and laws as especially fundamental and sacrosanct theories; but there is no distinction in logical form. All three are of the form of what (some) logicians have called strictly universal statements. A strictly universal statement purports to be true for all places and all times; for example, the statement 'All metals conduct electricity.' A strictly universal statement may, of course, be false, but that is an empirical, not a logical matter.
15 There are four such forces currently established: gravity, the 'strong' force that holds together the nucleus of the atom, the electromagnetic force, and the 'weak' force that governs radioactive decay. Magnetism and electricity were shown to be a single force by Maxwell in the nineteenth century; the electromagnetic force and the 'weak' force were successfully 'unified' by Glashow, Weinberg, and Salam in the 1960s. Their theory predicts the existence of entities that carry the force and these have been discovered experimentally. Many theories have been proposed unifying this 'electroweak' force with the 'strong' force, but none has so far been corroborated by experiment. The last step would be to unify such a force with gravity and theoreticians are already at work on this problem.
16 Said by Dumby in *Lady Windermere's Fan*, adding, 'Life would be very dull without them.' In fact, it would be impossible.
17 Reported in *Nature*, 10 June 1965: 121.

18 The historian-philosopher Thomas Kuhn has erected an entire sociology of science on the basis of this fact. He conceives of science normally being done within a framework of ruling ideas; a framework out of which it is impossible for any scientist to break short of a scientific revolution akin to a kind of religious conversion, which he calls a gestalt or 'paradigm' shift.

19 Wigner was a Hungarian *émigré* to the United States before Hitler's assumption of power in Germany. One of the founders of the atomic age, he was a Nobel Prize winner in physics.

20 For a series of technical arguments against the Copenhagen interpretation by Popper, see his 1982; 1966: 7–44; and his comments in Schilpp 1974: 72–6 and 1125–39. One of the most central objections that Popper has to the orthodox interpretation is that it has led many physicists to the mistaken belief that the probabilistic character of the theory must be explained by *a lack of knowledge* rather than by the statistical nature of the *problems* that quantum theory addresses. As he has said, 'the view that a probabilistic theory is the result of a lack of knowledge leads ... to the view that the probability of an event measures the degree of somebody's incomplete knowledge of [it]' (1966: 17).

21 A definitive account of the debate is to be found in Max Jammer's *The Philosophy of Quantum Mechanics* (1975).

22 See also his *In Search of Reality* (1985).

23 This is the orthodox Marxist view, but not all its advocates are Marxists. See for example, M.C. Robinson, 'Is Quantum Mechanics a Scientific Theory?' in the Marxist science journal *Science and Nature*, no. 6 (1983), 42–51. Robinson calls 'orthodox' quantum theory 'a computational tool' and a 'valuable set of cookbook recipes.' He argues that the list of physicists who have become disenchanted with the Copenhagen interpretation has been steadily growing. Popper (1982: 10) quotes Murray Gell-Mann, a leading contemporary theoretical physicist, as saying: 'Niels Bohr brainwashed a whole generation of theorists into thinking that the job [an adequate interpretation of quantum mechanics] was done 50 years ago.'

24 One of Weisskopf's reasons for this statement is that quantum mechanics is applied to the understanding of many things, ranging from the properties of metals to the processes in the interior of stars, all of which existed before humankind and therefore human consciousness evolved.

25 For a gentle satire on the notion that reality is brought into existence by the physicists' experiments, see Lewis Thomas, 'An Apology,' in his *The Medusa and the Snail* (1979).

26 See, for example, Richard Rorty, *Philosophy and the Mirror of Nature* (1980).

27 Our own galaxy, the Milky Way, contains millions of stars, many of which

must have solar systems like ours, and there are millions of galaxies other than ours, each containing millions of stars.

28 Proponents of the anthropic principle include Robert Dicke at Princeton University, who gave it its name, Brandon Carter, Roger Penrose, and Stephen Hawking at Cambridge University, and Bryce DeWitt and John Archibald Wheeler at the University of Texas at Austin. This section draws on Gale 1981 and Boslough 1985. For a fuller and more sophisticated account of the anthropic principle see Martin Gardner, 'WAP, SAP, PAP, and FAP,' *New York Review of Books*, 8 May 1986: 22–5.

29 These examples are taken from Boslough 1985: 122. The quotations from Hawking following are from p. 124.

30 Indeed, the originator of the 'many universes' conjecture, Hugh Everett of Princeton University, postulated that they must all be seen as equally real. His conjecture was an attempt to avoid certain difficulties inherent in the Heisenberg uncertainty relations that form part of quantum theory.

31 For example, it explains the stability of molecules, and hence, since the gene is a molecule, it explains how heredity is possible – how a certain family characteristic can be retained intact from generation to generation, in some cases over centuries (for example, the famous 'Hapsburg Lip,' a disfigurement of the lower lip peculiar to the Hapsburg dynasty).

32 Bohr himself crossed the threshold of metaphysics when his realization that, at the quantum level, the results depend on the state of the apparatus led him to formulate his idea of *complementarity*, an idea that took him far beyond quantum mechanics into a general philosophy of life. This was an amorphous and quasi-mystical doctrine that was never clearly worked out, but was influenced to some degree by his reading of his fellow-countryman Kierkegaard and the American philosopher William James (Holton 1973). When, in 1947, Bohr was asked to propose a design for his coat of arms, he placed the symbol for the Yin and the Yang at its centre. Another of the pioneers of quantum theory, Wolfgang Pauli, also developed mystical leanings, and collaborated with the psychoanalyst Carl Jung in a number of enterprises.

33 See his *The Tao of Physics* (1975).

34 And also some scientists, notably the physicist John Ziman. See his 1968 and 1978a.

35 A major exception being the American sociologist Robert Merton, who attempted to ground the objectivity of science in the peculiar nature of its practice; in particular, in the acceptance by scientists of a set of formal operating norms. See Merton 1973. Much in the modern sociology of science is directed to the refutation of Merton's ideas.

36 For an excellent account of recent sociology of science see Mulkay 1979.

CHAPTER 2

1 For a fuller account of the historical relationship between science and technology than can be given here, see Keller 1984. Keller's conclusion is similar to mine: science was not the prime originator of technology, or even the catalyst. Nevertheless, modern technology would be 'impossible without scientific training and comprehension of the nature of things' (Keller: 182).
2 The leading authority on science and technology in China is Joseph Needham. Part of his immense multi-volume history of science and technology in China has been abridged by Colin Ronan in Ronan and Needham 1978.
3 Before the introduction of the public clock (an innovation of immense social consequence) it was customary to divide the time between sunrise and sunrise into a fixed number of hours of daylight and an equal number of hours of darkness. This meant, of course, that the length of the daylight hour fluctuated according to the seasons (sundials were often calibrated to take account of this) and that a daylight hour was equal in length to a nocturnal hour only at the spring and autumn equinoxes. After the introduction of the public clock, time came to be seen as transcendent, as a uniform progress against which all events could be measured, rather than something (as we would now say) 'socially constructed.' As Isaac Barrow, Newton's predecessor in the Lucasian Chair of Mathematics at Cambridge, said: 'Whether things move or are still, whether we sleep or wake, Time pursues the even tenour of its way.'
4 I say 'with less veracity' because the reference that Harvey made to the water bellows was not in his great book on the circulation of the blood, but in a page of his lecture notes that was apparently written later. See G. Keynes, 1966: 182–3. So it may have been used simply as a striking image for public consumption.
5 For the origin of the term 'scientist' see Ross 1962. It was coined by the Reverend William Whewell, Master of Trinity College Cambridge, in 1834 and was much disliked by many nineteenth century British scientists including Thomas Henry Huxley. The term 'engineer,' however, dates (according to the *Oxford English Dictionary*) from at least the fourteenth century.
6 Quoted in Cohen 1976: 374
7 Mulkay 1979 presents a useful brief summary of some of this literature.
8 There is an old joke that most of the important discoveries in science are made while looking for something else. Other serendipitous scientific

discoveries that have had a technological impact are the discovery of the dangerous effects of halocarbons used as propellants in aerosol spray cans (made by investigators of the chemistry of the atmosphere of Venus) and (from the remoter past) the discovery of Joseph Priestley (1733–1804) that carbon dioxide (or, as he called it, 'fixed air') was soluble in water and agreeable to drink. He had discovered carbonated water, and thus became the father of the soft-drink industry. There is a parallel in technological innovation. Evidence suggests that many of the innovations coming from industrial laboratories resulted, not from a direct attack on the problem to be solved, but rather from work that was pointed in quite a different direction. For example, some of the early developments in radio came from people not in the communications industry; Kodachrome was invented by two musicians; and dacron was produced in the laboratories of a firm with no interest in the production of artificial fibres.

9 A 'tracer' is a substance used to follow the course of a physical, chemical, or biological process. Medawar says 'tracer techniques represent the most important advance in biochemical methodology in the present century, and it is no exaggeration to say that the whole of modern biochemistry and much of immunology is founded upon their use. An example in medicine is the use of radioactive iodine to examine the functions of the thyroid.' Entry 'Tracer' in Bullock and Stallybrass, *The Harper Dictionary of Modern Thought* (1977).

10 For a fascinating account of the utilization of pure science in the Bell Telephone Laboratories, which began in the 1930s, see Hoddeson 1980. The notion of 'basic' or 'fundamental' research may have originated here.

11 For another example of a Nobel Prize won by Bell employees see note 13.

12 Einstein can be regarded as a kind of 'limiting case' in the technology-science relationship. He is usually seen as the 'purest' of pure scientists, yet (in conjunction with Leo Szilard) he made a large number of patent applications, all but two of which were granted, for the design of a noiseless household refrigerator.

13 Charles Townes and Arthur Schawlow, Bell Laboratories, September 1957. Townes was a consultant to Bell Laboratories and Schawlow was a staff member. Another scientist who set the stage for the invention of the laser was Alfred Kastler, the discoverer of the optical 'pumping' effect.

14 Entry 'Symbiosis' in Bullock and Stallybrass 1977

15 Nevertheless, by the seventeenth century, great ingenuity in the design of simple instruments was evident. For example, the primitive, but ingenious microscope designed by Anthony van Leewenhoek late in the century en-

abled him to achieve a magnification of more than 250 diameters and to see and sketch bacteria, protozoa, sperm, and the capillaries of the blood system. The instrument consisted of no more than a tiny lens mounted in a hole in a strip of metal, with the specimen balanced on the point of a needle.

16 The address was reprinted in *Nature*, 30 June 1962. de Broglie's point has been made more light-heartedly by Hermann Bondi (1975: 7): 'The revolution in physics in the last two decades of the last century (what with the discovery of electrons and X-rays and so on) ... owes its particular timing in history entirely to the fact that not only did reasonably reliable vacuum pumps become available then for the first time, but also (and I believe equally important) that plasticine became available to stop the leaks.'

17 Galileo was one of the first of the moderns to develop his own instruments. He had a workshop and employed a professional instrument maker. It is said that he worked up a telescope capable of magnifying thirty times (the word 'telescope' appears to have been coined by Galileo in 1611). For the invention of the telescope see Van Helden 1977.

18 See Brooks 1967 for these and other examples.

19 Dewey's belief about the utilitarian origins of science contrasts curiously with many of the things he says elsewhere. For example, that the purpose of science should be *shifted* 'from contemplative enjoyment to active manipulation and control' (1930: 92); and that 'the glorification of "pure" science is a rationalisation of an escape; it marks the construction of an asylum of refuge, a shirking of responsibility' (1946: 144).

20 Benjamin Rumford (1753–1814) is, so far as I know, the only American to have been made a count of the Holy Roman Empire.

21 These 'superbolts' generate several thousand times the output of a large electricity generating station.

22 Some sociologists of science deny that scientific discoveries can entail their applications. Ben-David (1968: 50) argues that 'nothing is "implied" in a discovery beyond the questions answered by it, and those to which it is related by the traditions and mental habits of the people who are its prime consumers'; quoted by Barnes (1982b: 168), who adds: 'Scientific "discoveries" have no logical implications.' Ben-David's statement is true only if it is read as a tautology (i.e., if 'implied' is taken to mean 'seen now to be implied'). Barnes's statement is false. The fact that the implications of a scientific discovery go *unrecognized* (by the people who are its 'prime consumers') does not mean that no implications (I prefer 'entailments') exist. But Barnes does not believe in the possibility of objective knowledge.

23 This thesis has a long history. It has been by no means confined to the radical left. It is to be found also among the ideas of deeply conservative thinkers, especially in Germany, such as the philosopher and sociologist of knowledge Max Scheler and the philosophers Husserl and Heidegger.
24 See Marcuse's influential book *One-Dimensional Man* (1966).
25 'Internalist' history is so called because it tended to ignore the context in which the science of a specific time and place was done, and treated science as an insulated activity that was the product of purely individual intellectual interest. It was also the history of 'great' discoveries and 'great' scientists – or so its detractors say. Historians of science like George Sarton are labelled 'internalists.'
26 Internalist histories are said to be 'Whiggish' because they judged the science of a given epoch by the standards of a later age. The 'Whig interpretation of history' is a witty phrase we owe to Herbert Butterfield. It refers to the tendency of certain historians to see the 'present' (whenever that might be) as more enlightened than the past. The tendency is 'Whig' because it was a position adopted by some nineteenth century historians who *were* Whigs in politics – such as Thomas Babington Macaulay.
27 This is not to say that Galileo understood the nature of sunspots. He did not. He likened them to 'clouds, whips, and smokings.' But, as this striking phrase suggests, he had carefully observed their behaviour.
28 In the seventeenth century 'natural histories' of the Baconian sort were catalogues of facts. The natural histories of the early Royal Society, for example, dealt not only with comets but with iron-making and cloth-making; not only with the nature of fluids but also with mining. On occasion their investigations bordered on the ludicrous; for example, their report that the body of John Colet, dead for 160 years, proved to be 'of an ironish taste.'
29 Bacon's chief scientist in *The New Atlantis* is not a particularly attractive figure, but he is not unfamiliar either. He is the director of a research institute. He sits alone 'upon cushions' and 'has before him fifty Attendants, young Men all.'
30 Secularization was enshrined in the constitution of the Royal Society. One of its first secretaries, Henry Oldenburg, wrote in reply to a correspondent in 1663: 'The Royal Society says it is not its concern to have any knowledge of scholastic or theological matters ... These are the bounds to which the Royal Charter limits this British assembly of philosophers.'
31 One of the Marxist influences on those moderns who accept this identification was the extraordinary set of papers presented to the International Congress of the History of Science and Technology held in London, Eng-

land, in 1931. These papers (written by Soviet Marxists), and especially that on Newton by B. Hessen, have had a profound effect on 'externalist' historians of science. The papers are collected in *Science at the Cross Roads*, reprinted with an introduction by P.G. Werskey by Frank Cass Ltd., 1971. Werskey's account of the congress is well worth reading. Some of the participants whose essays appear in this book, including Hessen, were later 'liquidated' on Stalin's orders. See also L.R. Graham, 'The Socio-political Roots of Boris Hessen: Soviet Marxism and the History of Science,' *Social Studies of Science* 15, no. 4 (1985), 705–22.

32 For the idea that false theories may contain some truth (an idea that some people find very strange) see Popper 1969: 376–7 and 1972: 55–6. The same holds for any statement with empirical content. I take the following example from Iris Murdoch's novel *The Black Prince* (Penguin edition, p. 32):

> 'Is he – is this chap – a medical doctor?'
> 'Yes', I said. 'A friend of mine.'
> This untruth at least conveyed important information.

In fact the 'friend' is the narrator's brother-in-law and he has been struck off the register and therefore can no longer practise as a physician.

33 The native tribesman who chants while planting seed is not interested in testing the theory (that chanting promotes fertility); he is simply using it instrumentally.

34 This is not to deny that the practice of science *also* involves technical skill and practical knowledge. An experimental scientist must know how to use instruments and make them work, and may, even today, play a part in their design.

35 For increase in the *depth* of knowledge, see Popper 1983: 139–45.

CHAPTER 3

1 'When I open a new issue of a scientific journal, I do not scan the table of contents looking for an exciting novelty; on the contrary, I hope there will be nothing in it that I must read' (Salvador Luria 1984: 115). Luria is the discoverer of 'restriction enzymes,' for which he won a Nobel Prize, and is thus one of the founders of genetic engineering.

2 They have done this primarily to secure financial support for their research. Many scientists, however, are fascinated by the technological applications of their discoveries and wish to work on them.

3 For an excellent short account of the politics of science in the USSR, see Kneen 1984. On post-Maoist China, Kuhner 1984 is useful.
4 There is a large literature on this topic. See, especially, Joravsky 1970 and Medvedev 1969.
5 Soviet planning of science – or more strictly, attempts at planning – has become in fact more comprehensive since the 1960s.
6 The system of providing support through specific grants and contracts to specific individuals and projects has been sharply criticized by Roy and others.
7 For the British case see Shattock 1987.
8 Reported in the *New York Times*, 13 June 1982. The incident occurred when *Science*, the journal of the American Association for the Advancement of Science, disclosed that the Department of Agriculture was using a political test when appointing members to committees charged with judging the scientific merits of pure research proposals. Officials of the department were quoted as saying that, in choosing between scientists with comparable professional credentials, preference was given to those who were 'philosophically compatible' with the Reagan administration. An outcry from the scientific community succeeded in having the practice halted. Dr Keyworth's ideological position could be judged by his description of the press in America as 'drawn from a relatively narrow fringe element on the far left of our society' (*New York Times*, 24 March 1985).
9 There is a parallel in the Soviet Union. Since all Soviet research laboratories must produce a plan of future work (for incorporation in an overall scientific research plan, which in turn becomes part of the Five-Year Plan), it is common for a piece of research to be entered in the laboratory's plan only when it is already on the way to success.
10 Nietzche portrayed the Greek god Dionysus as 'liberated,' playful, and creative, and Apollo as sober. In Greek mythology Apollo is associated with temperance ('nothing in excess') and good sense. For more on this topic see the article 'On Conservative and Adventurous Styles of Scientific Research' by John Wettersten, *Minerva* 23, no. 4 (1985), 443–63.
11 Quoted in *Fortune* magazine, July 1977, p. 66.
12 Lewis Carroll, 'Fame's Penny Trumpet'
13 These words have a special poignancy when one considers the personal hardships to which Medvedev was subjected in the Soviet Union as a result of his independence of mind: loss of jobs, internal exile, a period of incarceration in a Soviet 'psychiatric' hospital, and finally, when he would not bend, expulsion from his homeland as an undesirable and loss of his Soviet citizenship. Medvedev's father, a professor of philosophy, perished

in a labour camp in 1941. We do not know what his 'crime' was. Zhores Medvedev's 'crime' was circulating, and later publishing in the West, truths about episodes in Soviet science that were unwelcome to the authorities.

14 The passions stirred by the conflict can be seen from the following passage written by a distinguished British scientist, Sir William Ramsay, in *Nature* in 1915, the year in which poison gas developed by German scientists was used for the first time at the second battle of Ypres: 'German ideals are infinitely far removed from the conception of the true man of science ... Germany must be bled white ... Will the progress of science be thereby retarded? I think not. The greatest advances in scientific thought have not been made by members of the German race ... So far as we can see at present, the restriction of the Teutons will relieve the world from a deluge of mediocrity. Much of their previous reputation has been due to Hebrews resident among them.'

15 For further details of this sordid episode see Haberer 1972: 715–20 and Schroeder-Gudehus 1973.

16 The laboratory in Montreal was code-named the Metallurgical Laboratory. That in Chicago was the laboratory headed by Enrico Fermi at which the reactor was built. The two scientists were Goldschmidt and Auger. The ban was lifted later in 1943 as part of the Quebec Agreement reached between President Roosevelt and the British prime minister, Winston Churchill.

17 The reluctance of some of the scientists of the past to publish openly – as against clandestinely within a narrow circle or sometimes in cipher – was largely due to fear of persecution by the state or religious authorities, which is perhaps what Bacon's scientists had in mind.

18 Fuchs was a German physicist who was later granted British citizenship. Originally a Social Democrat, he joined the Communist party while still a young man and felt the effects of Nazi persecution that befell most German communists in the 1930s. He later fled to Britain and worked for some time at the University of Edinburgh with Max Born. He was interned as an enemy alien after the fall of France, but was later released and returned to work with Born, and then at the University of Birmingham with Rudolf Peierls on the British atomic bomb project. He accompanied Peierls to Los Alamos when the latter went as part of the British contingent in the Manhattan District project.

19 The eminent Soviet scientist Peter Kapitza, who worked at the Cavendish Laboratory in England in the 1920s and early 1930s, returned to his homeland for what he supposed to be a visit in 1934 but, we now know, was detained on Stalin's orders. He remained there for the rest of his life.

In 1941, when the German armies were advancing on Moscow, he published openly (in *Isvestiia*) an appeal for Allied co-operation with the Soviet Union in making an atomic weapon.

20 It is true that Fuchs worked with Edward Teller at Los Alamos on theoretical problems relating to the 'Super,' or hydrogen bomb; but that project was so little advanced at that time that it is unlikely that his knowledge was of much assistance to Soviet scientists like Sakharov who much later worked on the Russian version of the hydrogen bomb.

21 Conceivably also to conceal the fact that Soviet geneticists are less advanced in some respect than their foreign 'competitors'

22 There is nothing new in principle about this, since it has been Soviet policy since at least the 1930s to acquire foreign technology and 'expertise' and adapt it to Soviet conditions. The problem today is posed by the military threat that the Soviet Union holds for the security of the Western Alliance. For a superb account of the difficulties surrounding attempts to control the flow of high technology see Macdonald 1986.

23 *Scientific Communication and National Security*, September 1982. Dr Dale Corson, a former president of Cornell University, was the chairman.

24 *United States of America v. The Progressive Inc.*, 1979. This case concerned the publication by an investigative journalist of the details of the construction of the hydrogen bomb. His article was put together from unclassified sources.

25 These and other examples are to be found in the 'Science and the Citizen' column in *Scientific American*, 251, no. 6 (December 1984).

26 An internal document of Harvard University concerning federal restrictions on the free flow of academic information and ideas (Shattuck 1984: 435). The Senate and House of Representatives remained deadlocked on this issue.

27 Executive Order 12356, April 1982

28 The California Institute of Technology, the Massachusetts Institute of Technology (both of which, in spite of their names, do much pure scientific research), and Stanford University in California

29 *Bulletin of the Atomic Scientists*, 41, no. 3 (March 1985)

30 It also denies it to those who speak for the public, and hence deprives the public of information that government agencies and the military find it inconvenient or embarrassing to reveal. For example, critics pointed out that the Department of Energy's category of 'controlled' but non-classified information could be used to suppress damaging information emerging from independent researchers about the effects of radiation from nuclear reactors.

31 For an informative treatment of the importance of the international migration of scientists in generating knowledge see Hoch 1987.

32 For a full treatment of American responses, see Lubrano 1981. It should be added that there are groups of scientists, and many scientific societies, that are concerned with the political persecution of scientists generally, for example, in other countries of the Soviet bloc, and in Latin America, and not merely with the better publicized Soviet dissidents. See Ziman 1978b and the report of a British study group published as *Scholarly Freedom and Human Rights*, 1977.

33 By contrast, the Royal Society of London has taken the more timid position that there exist other agencies of a more general character, such as Amnesty International, through which scientists should channel their protests.

34 For example, in my presence, by a speaker at a conference of the Society for Social Studies of Science held in Boston in 1977

35 When, after three years of defiance, the city fell (in 212 BC) the victorious Roman general ordered Archimedes spared, but he was killed by an overzealous Roman soldier.

36 Though Hitler made the mistake, later regretted, of sending large numbers of scientifically trained men to the front. The Nazis were largely indifferent to science and had little understanding of it.

37 I use the masculine form deliberately, since very few, if any, women scientists seem to work in closed weapons research laboratories.

38 Sutton found that the only peer reporting at Livermore for those engaged in weapons research was done in the course of annual program reviews sponsored by the Defense Nuclear Agency of the Department of Defense. Sutton's study is based on lengthy interviews with fourteen scientists drawn from five programs at Livermore: Nuclear Design and Testing, Controlled Thermonuclear Research (i.e., fusion for nuclear reactors), Lasers, Biomedical, and Energy. From internal evidence it would appear that the study is based on field-work done in the late 1970s.

39 This doctrine was first applied systematically in the Manhattan District project during the Second World War – except in the 'Technical Section' at Los Alamos, where it was successfully resisted by Robert Oppenheimer, the director of the laboratory. The 'need to know' doctrine means that workers in one segment of the organization can be told only what is necessary for them to do their assigned tasks.

40 It seems that this may occasionally be possible in the Soviet Union, though only for extremely eminent scientists. For example, it has been suggested that Kapitza refused to have anything to do with the Soviet hydrogen

bomb. If this is true, it was a singularly courageous act during the reign of Joseph Stalin.

41 The Treaty Banning Nuclear Weapons Tests in the Atmosphere, in Outer Space, and Under Water was signed by the United States, the Soviet Union, and Britain on 25 July 1963. Though other nuclear states remain non-signatories to the treaty, no tests have taken place in these environments since then.

CHAPTER 4

1 For instance, 'The happy seat of some new Race call'd Man,' *Paradise Lost*, II: 348

2 The widely accepted definition today of a species is 'a community of actually or potentially interbreeding organisms sharing a common gene pool.'

3 Useful studies are by John Haller 1971 and Nancy Stepan 1982.

4 Among these were his part in establishing the French Société d'Anthropologie (in 1859), his research on rickets and cancer, and (for which he is best remembered) his discovery of a small region in the left hemisphere of the brain connected with human speech known eponymously as Broca's Area. There is an interesting biography of Broca by Francis Schiller, *Paul Broca, Founder of French Anthropology, Explorer of the Brain*. See also an amusing essay by Carl Sagan, 'Broca's Brain' (in his book of the same title, 1979). Broca's brain was found by Sagan floating in formalin in the basement of the Musée de l'Homme in Paris.

5 Polygeny was one of the two opposed doctrines about the origins of the races of Man that were endlessly debated in the late eighteenth century and the first half of the nineteenth. Monogenists held that Man was descended from a single ancestral stock; polygenists held that each race had descended from different ancestral stocks. Polygeny tended to characterize American thinking on the question; monogeny, European thinking.

6 Actually, a very diverse people (only about 8000 now survive). The best known to the laity (in their Hollywood version) were the Comanche.

7 See also his chapter 13, 'Wide Hats and Narrow Minds,' an amusing account of a minor cause célèbre in which Broca was involved concerning the brain size of the great French anatomist Baron Georges Cuvier.

8 A strategem employed for much the same reasons by Arthur Jensen, as we shall see in the next section of this chapter

9 Quoted in Gould 1980a: 171

10 Buffon was a nominalist, that is, he held that universals have no real existence; only individuals exist. Classes of individuals may be given names,

but this is a conventional act. This means, in biology, that by *naming* groups species we *create* them as species. We 'call them into existence.'

11 *Essay on Man.* Compare also, for the late sixteenth century, Shakespeare's 'Take but degree away, untune that string / And hark! what discord follows'; and 'The heavens themselves, the planets, and this centre / Observe degree, priority, and place / Insisture, course, proportion, season, form / Office and custom, all in line of order'; *Troilus and Cressida*, I, iii, 109 and iii, 85, respectively. 'Insisture,' according to the *Oxford English Dictionary*, 'a word of obscure use in Shakspere [sic] ... perhaps [meaning] station.'

12 It can be argued that Darwin was at times influenced by his contemporaries against his better judgment, especially in the *Descent of Man*, some parts of which have led to fruitless debate about whether or not Darwin was a Social Darwinist. For example, he was led to agree with Francis Galton that natural selection could act on the moral faculties so that 'the best' moral faculties would survive and, as he put it, 'virtue will be triumphant.'

13 See, for example, Ghiselin 1969: 64–5.

14 One of his students, Paul Kammerer, an experimental naturalist and a strong believer in the inheritance of acquired characteristics, became notorious when it was found that one of his results, purporting to demonstrate Lamarckian inheritance, had been faked. He was later canonized by Lamarckians in the Soviet Union who alleged that he was the victim of a bourgeois plot to discredit him. He committed suicide, but it is not clear (as is frequently supposed) that this was due directly to his exposure as a fraud. Arthur Koestler wrote a book about him called *The Case of the Midwife Toad* (1971).

15 Quoted by Ernst Cassirer in his *The Problem of Knowledge* (1950), p. 162

16 This, and the following quotation are taken from Farley 1977: 76.

17 Ibid., pp. 80 and 81

18 See note 13. The paraphrase that follows relates to p. 123 of Ghiselin's book.

19 Gasman 1971, chapter 4

20 *Evolution and Ethics and Other Essays* (1920), 36–7

21 Many of Haeckel's ideas were adopted by the Nazis, or at least were entirely consistent with their thinking, and the party honoured Haeckel in 1934 on the 100th anniversary of his birth. Not all Monists, however, subscribed to protofascist ideology. Many indeed, as Gasman shows, rejected German nationalism and Volkish mysticism. Many were left-wing intellectuals, and some, even, were Marxists. The attraction of Monism for them was its opposition to religious dogmatism and metaphysical thinking. They were 'Machian' monists, in this respect, rather than Haeckelians.

22 It beguiled me, when I read it at the age of sixteen.
23 Quoted by Bernard Norton in his 1978a: 20–1
24 For an authoritative history of the eugenics movement see Kevles 1984.
25 The first occupant of the position was Karl Pearson, in 1911.
26 Galton's life and work have been surveyed in an excellent biography (Forrest 1974). Among Galton's many accomplishments not noted in the text were work in meteorology (he invented the term 'anticyclone' and introduced weather maps) and in the study of hallucinations. His more quirky enterprises included a proposed beauty map of Britain based on a classification of the women he met according to whether they were 'beautiful,' 'indifferent,' or 'ugly,' and a study of fidgeting among members of the Royal Geographical Society. A fellow spirit, Charles Babbage (1792–1871), an early pioneer in the invention of the computer, was also a passionate collector of 'facts.' On one occasion he wrote to the poet Tennyson suggesting that two lines from the poet's 'The Vision of Sin' should be changed. Tennyson had written: 'Every minute dies a man, / Every minute one is born.' Babbage's suggested amendment was 'Every minute dies a man, / And one and a sixteenth is born.' He apparently added, 'the exact figures are 1.167, but something must, of course, be conceded to the laws of scansion.' It is not clear whether he had his tongue in his cheek when he wrote this.
27 'Inherited' in the Darwinian sense; Lamarckian inheritance would have led to 'environmentalism' (that is, to improvement of inheritable qualities by altering the environment). Galton was a 'hereditarian.'
28 'The Possible Improvement of the Human Breed under the Existing Conditions of Law and Sentiment,' reprinted in part in Abrams 1968: 260–4
29 For the eugenics movement in Britain see Mackenzie 1976 and 1981, and Farrall 1970.
30 There is an excellent recent biography of Ellis by Phyllis Grosskurth (*Havelock Ellis*, 1980). A reviewer of this book in the *London Review of Books*, 3–16 July 1980, p. 21, describes Ellis as 'a man much given to the sexless discussion of sex who made the world safe for Masters and Johnson.'
31 Sir Keith Joseph
32 Not that this concern was entirely absent in the British case, for there had been a large influx of immigrants from Eastern Europe into the slums of London in the 1880s and 1890s.
33 For a fuller account, see Kimmelman 1983.
34 Title of his address to the National Farmers' Congress in 1912. Quoted by Kimmelman 1983: 186
35 Reporting on a new study of Dugdale and the Jukes by E.A. Carlson

36 Another example of a scandalous abuse of evidence was uncovered by Stephen Jay Gould and is reported in his 1981a: 168–73. It concerns the equally famous 'Kallikak' family discovered by H.H. Goddard (1866–1957), who invented the term 'moron' and divided the mentally defective into a hierarchy of types. The family is said to have had two lines of descent: one line, of 'degenerates,' coming from the union of a soldier of the Revolutionary War ('Martin Kallikak') with a 'feeble-minded' tavern girl; the other line, of upright 'American types,' from the marriage of the same man to a Quaker woman. This demonstration of the power of heredity was widely used by eugenicists, and it was still appearing in text-books in the 1950s (see Block and Dworkin 1977: 373). Goddard's book about the family contains photographs of some of the descendants that he found living in poverty in the pine barrens of New Jersey. Gould discovered that these had been retouched in order to make the subjects look depraved.

37 Recent developments in biological science are opening up sophisticated aspects of negative eugenics of which the early eugenicists never dreamt. Modern genetic engineering holds out the hope of eliminating inheritable genetic defects, and already certain congenital ailments like phenylketonuria could be eliminated altogether by identifying the carriers and dissuading them from marrying. Since many of these defects are relatively rare it would no doubt, in principle, be possible administratively to make treatment or abstention from marriage compulsory. Old-style eugenics may be dead, but the underlying moral issues (e.g., is there a right to give birth to defective children?) remain with us.

38 Moreover, *moral* scruples might sometimes affect eugenic policies even in highly regimented totalitarian regimes. Hitler's (negative eugenic) 'Euthanasia' program, which he personally authorized in September 1939, was terminated two years later (though not until it had claimed the lives of some 90,000 mentally defective and chronically ill people) by reason of a public outcry that has been described as the only serious civilian protest in the history of the Third Reich. It has been said that this is the reason the extermination policies of the regime (the 'Final Solution') were carried out in extreme secrecy, with elaborate camouflage, and by the SS, who could be counted on to give unconditional obedience to orders. (See Craig: 1983.)

39 For a masterly summary see Block and Dworkin (1977: 410–540).

40 Karl Pearson was probably the first to attach a percentage number to the inheritance of intelligence. His 'estimate' was 90 per cent!

41 One of my former graduate students who was present in February 1977

when Eysenck attempted to speak at the University of Leeds wrote: 'Emotions were strong at that meeting and one was almost overwhelmed by the seeming irrationality at work – until one realised that at the same time there was a strong reason behind it. Eysenck did not give the impression that he was a scientist who was naive about the political implications of his work. When asked if he would publicly abhor the fact that some of his work was being quoted in propaganda by the National Front [a British Fascist organization, and a direct descendant of the pre-war British Union of Fascists] he refused to answer or to accept any responsibility, saying he was just a scientist.' Eysenck is a somewhat ambiguous figure in the company of persons like Shockley. Some of his popular books like *Uses and Abuses of Psychology*, published by Penguin Books in 1953, are very moderate in tone, and the chapter 'What Do Intelligence Tests Really Measure?' makes few of the large claims one would expect from reading some of his later utterances.

42 A dominating influence on the British educational system between the wars. See this section, pp. 106–7, for more on Burt.

43 Quotations are from an interview with Daniel Yerkin in the *Times Higher Educational Supplement*, 24 May 1974, and an interview with David Dickson in the *New Scientist*, 22 February 1973. As to the charge of Lysenkoism, if there are such echoes in this debate, it is more appropriate to attribute them to the Jensenites. Jensen himself, in his long, exculpatory preface to *Genetics and Education* (1972), accuses his critics of 'thinking incorrectly genetically.' This is precisely what Lysenko said of *his* critics.

44 See *Encounter* 39, no. 6 (1972).

45 *Guardian*, London, 10 May 1973, quoted in Rose and Rose, 1976: 114. Eysenck is not, of course, a geneticist and does not do research in genetics, though he makes frequent pronouncements about it.

46 The more extreme version of this argument is whether 'sensitive' research should be done at all.

47 Malitza 1979 reports that a researcher in biophysics once told him that an evaluation of a piece of his work by a review committee began with the words, 'Your seriousness is proved by the fact that you did not speak to the press'! The *New England Journal of Medicine* is one learned journal that will not accept articles reporting research results that have been previously released to the press or broadcast on radio or television.

48 A reviewer of Jensen's book *Genetics and Education* (1972) wrote that a latter-day Brecht would be unlikely to write a new *Galileo* with Jensen as his hero. Predictably, he was only half right. No latter-day Brecht, but a sociologist of science (Barnes 1974: 131–6) has done just that. Barnes examines Jensen's credentials and finds that he is a scientist 'within the

normal pattern of practice,' that his 1969 paper is a good paper 'as scientific papers go,' although it is 'heavily polluted' with policy considerations, and that it offers no grounds for contending 'that particular interests or social convictions were influencing his judgment.' He was, 'taken out of context,' disinterestingly pursuing the truth. Like Galileo, he was opposed by the current orthodoxy, by the standards of which he was wrong. What his opponents were doing was what Galileo's opponents were doing: defending an occupational privilege. These highly idiosyncratic notions and the convoluted logic by which they were reached can best be exposed by considering the substantive issues discussed in the text of this chapter.

49 I use the labels 'Jensenism,' 'Jensenite,' and 'Jensenist' throughout simply to distinguish a particular brand of IQ psychology associated with Jensen, Herrnstein, Eysenck, et al. and their intellectual forerunners like Terman, Spearman, and Burt. There are many IQ psychologists who repudiate Jensenism and whose theoretical and experimental work refutes it at almost every point.

50 We are assured that Sir Cyril Burt, for one, did not believe that it could be measured directly, though perhaps Eysenck thinks it can be (for example by EEG tests).

51 More specifically, that intelligence is a relatively fixed attribute that sets an upper limit on individual intellectual functioning, leaving little room for cognitive development

52 Eysenck once suggested that a possible reason for the 'finding' that Irish schoolchildren score poorly on IQ tests by comparison with English children is that the more intelligent Irish emigrated to North America.

53 Good examples are Medawar (1977b), Gould (1980b), Lewontin (1981), and especially Layzer (1977).

54 Rose (1976: 116n) has drawn attention to the fact that Eysenck sometimes refers to intelligence as a concept (i.e., mental construct) rather than as a thing; but as Rose comments, 'it is difficult to see how a concept can have its hereditability estimated.'

55 Jensen has argued, in *Bias in Mental Testing*, that 'the constructors, publishers and users' of IQ tests can 'remain agnostic' about causes.

56 *Times Higher Educational Supplement*, 27 September 1974

57 Philip Morrison in a review of David Hawkins's *The Science and Ethics of Equality* has pointed out that the WISC test scores as 'wrong' the answer given by a New Mexican Indian child to the question 'Who discovered America?' The child's answer was 'We did.'

58 The difference between two races is a difference between averages and reflects differences in the statistical frequency of a particular trait.

59 Quoted from News and Comment, 'IQ and Heredity: Suspicion of Fraud

Beclouds Classic Experiment,' *Science* 194, no. 4268 (28 November 1976), 916–19; written by Nicholas Wade

60 Burt's official biography, which includes an assessment of this sad story, is by Hearnshaw (1979). See also Gould 1981a: 234–320. As mentioned in chapter 5, in his declining years Burt also wrote quite extensively on extra-sensory perception and psychical research. See, for example, Burt 1975.

61 Jensen's argument that changing the child's environment can have very little effect in raising the IQ score is refuted in a study by a group of French educational psychologists, reported in *Scientific American* 47, no. 6 (December 1982), which shows that, in their sample of adopted children, it was raised by as much as 14 points.

62 Operationalism was an important element also in the thinking of the early logical positivists such as Otto Neurath.

63 Logically speaking the weakest hypothesis is that which is not falsifiable, that is, the hypothesis that says nothing. In this case it would be: 'Differences in IQ are either entirely due to hereditary factors or they are entirely due to environmental factors.' The method of the null hypothesis has been said to encourage researchers to seek to answer those questions that are most likely to yield confirmations, and to avoid improving experimental and observational techniques that are likely to increase the possibility of yielding falsifications.

64 The empirical objection to this Jensenist argument is that although each individual environmental factor may be only weakly determinative of intelligence, in combination they may exercise a powerful effect. Jensen assumes that if all the possible environmental factors that can be thought of are eliminated one by one the environmentalist thesis is 'disproved'; but this does not follow.

65 This is not a new idea. We find at least its seed in Thomas Henry Huxley's remark (he was attacking Social Darwinism) that 'the evolution of society is a process of an entirely different kind from that which brings about the evolution of species'; and also in Marx's contradiction of Feuerbach's materialism – that Man is a part of Nature, but he changes Nature, and by so doing changes himself. According to Medawar, the first biologist to clearly recognize the Lamarckian character of cultural evolution was T.H. Morgan, one of the pioneers of modern genetics.

66 Indeed, he argued in *On Human Nature* (1978) that in entirely genetically driven animal societies like insect societies 'the sterile workers are ... more cooperative and altruistic than people,' and that 'in a broad sense' insect societies possess 'a higher morality' than we do (pp. 22–3).

67 The Library of Congress cataloguing data show the subject of *On Human Nature* as '1. Sociology 2. Social Darwinism'; obviously the work of a critic.
68 Some of it conveniently presented in three collections of 'readings': A.L. Caplan, ed., *The Sociobiology Debate* (New York: Harper and Row – Colophon Books 1978); Ashley Montagu, ed., *Sociobiology Examined* (Oxford: Oxford University Press 1980); and Michael Gregory, Anita Silvers, and Diane Sutch, eds., *Sociobiology and Human Nature* (San Francisco; Jossey-Bass Publishers 1978). One of the most recent, and definitive, treatments by a single author is Philip Kitcher, *Vaulting Ambition: Sociobiology and the Quest for Human Nature* (Cambridge, Mass.: The M.I.T. Press 1985). This book is not directed solely at Wilson. See also Roger Trigg, *The Shaping of Man: Philosophical Aspects of Sociobiology* (New York, NY; Schocken Books 1983); Michael Ruse, *Sociobiology: Sense or Nonsense?* (Dordrecht, Holland: D. Reidel 1979); and The Dialectics of Biology Group, Steven Rose, ed., *Against Biological Determinism*, and *Towards a Liberatory Biology* (New York, NY: Allison and Busby, distributed Schocken Books 1982).
69 Much of this article appears again in two chapters of *On Human Nature*.
70 Correspondence between Wilson and Medawar in the *New York Review of Books*, 31 March 1977
71 In so far as it is possible for anyone to be a determinist without falling into paradox; for as Epicurus wrote: 'He who says that all things happen of necessity cannot criticise another who says that not all things happen of necessity. For he has to admit that this assertion also happens of necessity.'
72 The dust-jacket on *The Selfish Gene* reads 'Thus Richard Dawkins introduces us to ourselves as we really are – throwaway survival machines for our immortal genes. Man is a gene machine, a robot vehicle, blindly programmed to preserve its selfish genes.'
73 See chapter 1, p. 20.
74 But not all such acts. For example, some birds expose themselves to danger by feigning injury in order to divert the attention of a predator from the fledglings in the nest. In cases such as this there is no mystery; the birds are protecting their genes in a very clear way.
75 A fascinating example of 'technical complexity in detail' not involving sacrifice, but apparent 'altruism' nevertheless, is provided by the chimpanzee. It has been observed that many male chimpanzees (often including dominant males) refrain from mating, even though the number of males and females in a group or tribe is roughly equal, and in spite of the fact that reproductive opportunities are limited because female chimpanzees are sexually receptive for only a few weeks every five years (which means,

in psychic phenomena), and Rhine himself. On Cyril Burt, see chapter 4, p. 20.

17 Instances of fraud in science are surveyed in Broad and Wade's controversial *Betrayers of the Truth: Fraud and Deceit in the Halls of Science* (1983). Reviewing the book in the *London Review of Books* (17–30 November 1983, pp. 5–7), Sir Peter Medawar suggested that 'Rather than marvel at, and pull long faces about, the frauds in science that have been uncovered, we should perhaps marvel at the prevalence of, and the importance nowadays attached to, telling the truth – which is something of an innovation in cultural history.'

18 Joseph Hixon's *The Patchwork Mouse* (1975) is a good journalistic account of the case. See also a review by Sir Peter Medawar in the *New York Review of Books*, 15 April 1976. Medawar was a consultant in the Summerlin case.

19 Modern parapsychology makes increasing use of sophisticated technology such as computers and random-number generators, all of which give a spurious air of 'scientificity' to the enterprise and are supposed to reduce the possibility of human error. Levy's electric shocks were administered to the rats by an automatic device that turned the current on and off at random intervals. Hence the claim was that, if the current were turned on more often than the laws of probability allow, this must be due to the rats' psychokinetic powers. What Levy did was to repeatedly pull the plug on the recording device in order to produce results better than chance.

20 This story has been told many times, for example, by Derek de Solla Price (1962: 84–9), Rostand (1960: 13–29), and Judson (1980: 172–3), but the best recent account is Klotz (1980: 168).

21 Gardner 1981 contains many other instances. The entire subject has been authoritatively reviewed by the psychologist C.E.M. Hansel, *ESP and Parapsychology*, (1980). See also D. Marks and R. Kammann, *The Psychology of the Psychic* (1980); T. Hines, *Pseudoscience and the Paranormal* (1987); and H. Gordon, *Extrasensory Deception* (1987).

CHAPTER 6

1 The neglect of science shown by the philosophers who founded the tradition of linguistic philosophy was clearly expressed by Frank Ramsey, a Cambridge logician much influenced by his contemporary Wittgenstein. Ramsey wrote: 'I don't feel the least humble before the vastness of the heavens ... My picture of the world is drawn in perspective and not like a model to scale. The foreground is occupied by human beings and the stars are as small as threepenny bits. I don't really believe in astronomy, except

our psyches and, among other things, that modern warfare is an abreactive re-enactment of these traumatic events.

8 For further evidence of Velikovsky's erroneous ideas see informative articles by C. Leroy Ellenberger appearing in *Kronos*, a journal devoted to Velikovsky studies. Dr Ellenberger is a former senior editor of that journal.

9 For this, see especially his address to the AAAS symposium, reprinted in revised form in the *Humanist*, Nov.–Dec. 1977. This contains clear evidences of megalomania. His posthumously published defence, *Stargazers and Gravediggers* (Velikovsky 1983), appeared under the imprint of the William Morrow Company, not Doubleday. It relates all his grievances (some indisputably justified), but continues to show no understanding at all of what the scientific objections to his ideas are really about.

10 Parapsychology as an academic discipline has by no means displaced the older bodies concerned with 'psychic research.' The Society for Psychical Research, as mentioned in the text, still exists, as does the American Society for Psychical Research, which was founded shortly after its British counterpart. Some university-based parapsychologists are members of psychical societies and contribute to their proceedings. The Parapsychology Foundation in New York City was founded by a spirit medium. The early history of psychic research in Britain is surveyed in a book by John Cerullo (1982). The early history of parapsychology under Rhine is traced in Mauskopf and McVaugh 1980; see also Oppenheim 1985.

11 About two-thirds of the members of the Parapsychological Association have a PhD. Only about one-half of the members work in the United States.

12 Documented in an article that appeared in the *New York Review of Books*, 17 May 1979, and reprinted in Gardner 1981: 185–206

13 For example, an experiment to measure the displacement of radiowaves from a quasar (3C 279); measurements of the deflection of radio signals from the spacecraft Mariner 6 and Mariner 7 while in the sun's gravitational field in 1970. The fact that confirmations never establish a theory as certainly true can be seen from the fact that the experimental results are also consistent with the prediction of the degree of deflection that is derived from a different theory (the 'scalar-tensor' theory). Nonetheless, many other predictions of Einstein's have also been confirmed. But it is highly unlikely that Einstein's general theory will never be superseded. See Clifford Will's excellent *Was Einstein Right?* (1986) for a comprehensive treatment.

14 For fuller details of this and other cases, see Gardner 1981, and other works cited in note 21.

15 For a discussion of 'the fraud hypothesis,' see Pinch 1979, Hardin 1981, and Pinch and Collins 1981.

16 For example, the philosopher C.D. Broad, Sir Cyril Burt (both believers

in psychic phenomena), and Rhine himself. On Cyril Burt, see chapter 4, p. 20.

17 Instances of fraud in science are surveyed in Broad and Wade's controversial *Betrayers of the Truth: Fraud and Deceit in the Halls of Science* (1983). Reviewing the book in the *London Review of Books* (17–30 November 1983, pp. 5–7), Sir Peter Medawar suggested that 'Rather than marvel at, and pull long faces about, the frauds in science that have been uncovered, we should perhaps marvel at the prevalence of, and the importance nowadays attached to, telling the truth – which is something of an innovation in cultural history.'

18 Joseph Hixon's *The Patchwork Mouse* (1975) is a good journalistic account of the case. See also a review by Sir Peter Medawar in the *New York Review of Books*, 15 April 1976. Medawar was a consultant in the Summerlin case.

19 Modern parapsychology makes increasing use of sophisticated technology such as computers and random-number generators, all of which give a spurious air of 'scientificity' to the enterprise and are supposed to reduce the possibility of human error. Levy's electric shocks were administered to the rats by an automatic device that turned the current on and off at random intervals. Hence the claim was that, if the current were turned on more often than the laws of probability allow, this must be due to the rats' psychokinetic powers. What Levy did was to repeatedly pull the plug on the recording device in order to produce results better than chance.

20 This story has been told many times, for example, by Derek de Solla Price (1962: 84–9), Rostand (1960: 13–29), and Judson (1980: 172–3), but the best recent account is Klotz (1980: 168).

21 Gardner 1981 contains many other instances. The entire subject has been authoritatively reviewed by the psychologist C.E.M. Hansel, *ESP and Parapsychology*, (1980). See also D. Marks and R. Kammann, *The Psychology of the Psychic* (1980); T. Hines, *Pseudoscience and the Paranormal* (1987); and H. Gordon, *Extrasensory Deception* (1987).

CHAPTER 6

1 The neglect of science shown by the philosophers who founded the tradition of linguistic philosophy was clearly expressed by Frank Ramsey, a Cambridge logician much influenced by his contemporary Wittgenstein. Ramsey wrote: 'I don't feel the least humble before the vastness of the heavens ... My picture of the world is drawn in perspective and not like a model to scale. The foreground is occupied by human beings and the stars are as small as threepenny bits. I don't really believe in astronomy, except

67 The Library of Congress cataloguing data show the subject of *On Human Nature* as '1. Sociology 2. Social Darwinism'; obviously the work of a critic.
68 Some of it conveniently presented in three collections of 'readings': A.L. Caplan, ed., *The Sociobiology Debate* (New York: Harper and Row – Colophon Books 1978); Ashley Montagu, ed., *Sociobiology Examined* (Oxford: Oxford University Press 1980); and Michael Gregory, Anita Silvers, and Diane Sutch, eds., *Sociobiology and Human Nature* (San Francisco; Jossey-Bass Publishers 1978). One of the most recent, and definitive, treatments by a single author is Philip Kitcher, *Vaulting Ambition: Sociobiology and the Quest for Human Nature* (Cambridge, Mass.: The M.I.T. Press 1985). This book is not directed solely at Wilson. See also Roger Trigg, *The Shaping of Man: Philosophical Aspects of Sociobiology* (New York, NY; Schocken Books 1983); Michael Ruse, *Sociobiology: Sense or Nonsense?* (Dordrecht, Holland: D. Reidel 1979); and The Dialectics of Biology Group, Steven Rose, ed., *Against Biological Determinism*, and *Towards a Liberatory Biology* (New York, NY: Allison and Busby, distributed Schocken Books 1982).
69 Much of this article appears again in two chapters of *On Human Nature*.
70 Correspondence between Wilson and Medawar in the *New York Review of Books*, 31 March 1977
71 In so far as it is possible for anyone to be a determinist without falling into paradox; for as Epicurus wrote: 'He who says that all things happen of necessity cannot criticise another who says that not all things happen of necessity. For he has to admit that this assertion also happens of necessity.'
72 The dust-jacket on *The Selfish Gene* reads 'Thus Richard Dawkins introduces us to ourselves as we really are – throwaway survival machines for our immortal genes. Man is a gene machine, a robot vehicle, blindly programmed to preserve its selfish genes.'
73 See chapter 1, p. 20.
74 But not all such acts. For example, some birds expose themselves to danger by feigning injury in order to divert the attention of a predator from the fledglings in the nest. In cases such as this there is no mystery; the birds are protecting their genes in a very clear way.
75 A fascinating example of 'technical complexity in detail' not involving sacrifice, but apparent 'altruism' nevertheless, is provided by the chimpanzee. It has been observed that many male chimpanzees (often including dominant males) refrain from mating, even though the number of males and females in a group or tribe is roughly equal, and in spite of the fact that reproductive opportunities are limited because female chimpanzees are sexually receptive for only a few weeks every five years (which means,

since most groups contain about fifteen males and fifteen females, that, on average, only three females will be receptive each year and only three males will be able to sire offspring in that year). Yet the pressure of natural selection favours behaviour that maximizes opportunities to reproduce. The answer to this puzzle appears to be that the females of the group come originally from another territory whereas the males are all closely related genetically because they are descended from the same patriarchal line. Thus their relatedness explains their apparent 'altruism.' They cannot all successfully mate, but some of their brothers and cousins can.

76 Except the bluenosed dolphin

77 Much information is genetic and inherited: how to react, to anticipate, to learn, etc. Some recent research has suggested that trees under attack by insects may inform (warn by chemical signalling) healthy neighbours so that they may release inhibiting chemicals. Even bacteria can communicate.

CHAPTER 5

1 Lavoisier was one of the founders of modern chemistry and a discrediter of the phlogiston theory. He was also a tax collector, for which he was guillotined during the French Revolution, thus cutting off his life at the early age of fifty-one.

2 By de Vries, Correns, and Tschermak. For a full and authoritative account see Orel 1984. See also Brannigan 1979. Brannigan quotes Sir Ronald Fisher as saying that each generation found in Mendel's papers only what it expected to find and therefore ignored what did not confirm its own experience.

3 For an extremely interesting account of the background to the Piltdown affair see Hammond 1982.

4 When Martin Gardner went with a group of scientists to remonstrate with officials of one of the major American television networks concerning the network's 'outrageous pseudodocumentaries about the marvels of occultism,' one shouted in anger: 'I'll produce anything that gets high ratings.' Gardner remarks: 'I thought to myself, this should be engraved on his tombstone' (Gardner 1981).

5 For an early attack on Rorvik see Goodell and Goodfield 1978.

6 Stuart Sutherland in the *Times Literary Supplement*, 2 January 1981, p. 12

7 In a later work, *Mankind in Amnesia*, subtitled 'An Enquiry into the Future of the Human Race,' Velikovsky returned to his psychoanalytic and Jungian theme, claiming that our memories of the global catastrophes he had ascribed to physical events in the solar system are deeply buried in

as a complicated description of part of the course of human and possibly animal sensation.' (Quoted by J.M. Keynes in his memoir of Frank Ramsey, Keynes 1972: 345. Ramsey died in 1929 at the age of 26.) For an interesting paper on this topic see Bloor, 'Are philosophers averse to science?' and a response by Briskman, both in Edge and Wolfe 1973. A good example of what I mean by contemptuous indifference is provided by the greatly respected modern philosopher Donald Davidson in the following passage: 'Well, this earth of ours is part of the solar system, a system partly identified by the fact that it is a gaggle of large cool, solid bodies circling round a very large hot star' ('Thought and Talk' in S. Guttenplan, ed., *Mind and Language* [Oxford: Oxford University Press 1975]). Professor Davidson must surely know that many of the planets are neither cool nor solid; and his reference to a 'gaggle' of planets, a word usually limited to a flock of geese, is undeniably patronizing.

2 I would wish to avoid the terms 'humanist' and 'humanism' if that were possible – if for no other reason than that they beg the question, putting science on one side of a fence and the arts and letters on the other. My objection is not directly related to the notion of the 'Two Cultures' – the 'humanistic' and the 'scientific' – a theme developed by C.P. Snow (1959) that led to an unedifying discourse with the Cambridge literary critic F.R. Leavis. It relates rather to my contention that science *is* a humanity, and that scientists should look upon themselves (as the best of them already do) *as* humanists. Unfortunately most budding scientists are now trained to believe otherwise, and this is a tragedy. For an elegant and subtle discussion of the idea of the two cultures see Sir Peter Medawar's Romanes Lecture, 'Science and Literature,' reprinted in *The Hope of Progress* (1972).

3 Ecclesiastes, 1:18

4 Wordsworth's generally hostile view of science softened in later life as his moving epitaph on Newton in the second version of *The Prelude* makes clear: 'Where the statue stood / Of Newton with his prism and silent face, / The marble index of a mind for ever / Voyaging through strange seas of thought, alone.' The second version of *The Prelude* was published some forty-five years after the first was completed. Neither appeared in Wordsworth's lifetime.

5 Eccentric exceptions can always be found, of course – for example the physicist I.I. Rabi, who once described the arts as 'not the kind of thing that will inspire men to push on to new heights,' and the works of Shakespeare as 'wonderful, glorified gossip.' But, on the whole, I think there are fewer absurd statements by scientists about the arts than there are by 'humanists' about science.

6 Cardinal Bellarmine. The cardinal wrote in a letter dated 4 April 1615 (not directly to Galileo but to a Carmelite monk, Foscarini, who had published a book in defence of Galileo and Copernicus): 'Your Reverence and Signor Galileo act prudently when you content yourselves with speaking hypothetically ... For to say that the assumption that the Earth moves and the Sun stands still saves all the celestial appearances better than [does the Ptolemaic system] is to speak with excellent good sense and to run no risk whatever. Such a manner of speaking suffices for a mathematician. But to want to affirm that the Sun, in very truth, is at the centre of the universe ... and that the Earth ... revolves very swiftly round the Sun, is a very dangerous attitude and one calculated ... to injure our holy faith by contradicting the Scriptures.' Bellarmine, an able scholar and a humane man, was a major influence in the Catholic Counter-Reformation. He was concerned to defend the Faith in a time of great tribulation for the Church of Rome. Galileo was the victim of religious politics. A recent book by the Italian scholar Pietro Redondi (1987) argues that Galileo's adherence to atomism was a major factor in his downfall, indeed that his Copernicanism was not the real reason for it. The fear on the part of the Holy Office, Redondi claims, was that atomism could be used by Protestant reformists to attack the church's doctrine of transubstantiation – the miraculous changing of the bread and wine into the body and blood of Christ (i.e., the real presence of Christ in the Sacrament). It is clear that, if Mr Redondi is correct, the issue was much more complex than history has previously recorded. The fact remains, however, that instrumentalist arguments *were* used against the Copernican theory and that Galileo was not an instrumentalist. He rejected the view voiced by the cardinal.

7 The Church of Rome apologized, in effect, to Galileo in 1965, 332 years after his public recantation, when Pope Paul VI called him 'a great spirit' and a man 'of immortal memory.' A commission to study 'the Galileo affair' was appointed by the church in 1979 but has yet to render its report.

8 My own university has had a visitation from one of the leading 'creationists' (October 1982) and I have had an exchange with a correspondent belonging to a body called Creation Science of Ontario.

9 As the sociologist Ernest Gellner ironically says, 'In the 18th and 19th centuries atheism was taught by Rationalists, but now it seems mainly the province of theologians' (1974: 33).

10 'There is talk of a new astrologer who wants to prove that the earth moves ... But that is how things are nowadays: when a man wishes to be clever he must needs invent something special ... The fool wants to turn the whole art of astronomy upside down. However, as Holy Scripture tells us, so did

Joshua bid the sun to stand still and not the earth'; Martin Luther, quoted in Koestler 1964: 572.

11 On the accommodation of Catholic theology to science in France in the latter part of the nineteenth century, see Paul 1979. A leading figure in the rapprochement was the physicist and Catholic philosopher of science Pierre Duhem. Duhem has exerted a notable influence on modern philosophy of science, yet his response to the challenge of science is, at bottom, that traditionally associated with Catholicism: scientific theories are neither true nor false but merely appropriate and useful instruments for describing and calculating aspects of reality. Thus, science and religion are autonomous domains and nothing is to be gained by their coming into conflict. Duhem wrote in his 'Physics of a Believer,' published in the *Annals of Christian Philosophy* in 1905, that the task was to define physics in such a way that 'positivists and metaphysicians, materialists and spiritualists, nonbelievers and Christians might work with common accord.' He regarded his philosophy of science as eliminating 'the alleged objections of physical science to metaphysics and the Catholic faith.'

12 The literature is immense. For a concise and interesting essay see Young 1985.

13 That is, mainly British and North American scientists. French biologists were mostly Lamarckians, and we have seen what happened to evolution in Germany as propagated by Haeckel (see chapter 4).

14 For example (on inductivist grounds), that the theory 'went beyond the facts'

15 A distinction drawn by P.M. Rattansi in his very interesting essay 'Science and Religion in the 17th Century' in Crosland 1975

16 A mistake, incidentally, that is connected with the false idea that scientific knowledge can be demonstrably certain – which is one reason why I say the belief that science entails adherence to atheism is 'positivistic.'

17 There is now a considerable literature on creationism and several excellent critiques of it. This being so, I provide only a sketch of it here. Interested readers are referred to Nelkin 1977a, 1977b, and 1982; Newell 1982; Kitcher 1982; Ruse 1982; Cloud 1977; Gould 1983; a symposium on creationism in *Science, Technology and Human Values* 7, no. 40, summer 1982; McGowan 1984.

18 Bryan died from a stroke five days after the trial ended. The trial, before the Supreme Court of Tennessee, is the subject of a play, *Inherit the Wind*, first produced on Broadway in 1955.

19 Engineers tend to be prominent in the creationist movement. When interviewed by Nelkin (1982: 87) some expressed ambivalence about nuclear

technology because the techniques for waste disposal were 'based on evolutionary assumptions.' No doubt this refers to the proposal for the long-term disposal of nuclear waste by deep burial in stable rock formations that are several hundred million years old. If the earth were created only some 10,000 years ago – as creationists assert – creationist engineers would be right to be worried.

20 Reported in the *New York Times*, Sunday 29 August 1982

21 She was 'reborn' while undergoing a Pentecostal baptism following a traumatic personal experience.

22 The title continues: 'to protect academic freedom by providing student choice; to ensure freedom of religious exercise; to guarantee freedom of belief and speech; to *prevent* establishment of religion; *to prohibit religious instruction concerning origins*; and to bar discrimination on the basis of creationist or evolutionist belief' (emphasis added). This title indicates the care taken to close every loophole through which the opponents of creationism might climb; it also reflects the varying legal judgments on the basis of which creationists had lost in previous cases.

23 *New York Times*, Sunday 22 March 1981

24 The physicist Richard Feynman amusingly describes this tactic: 'Some years ago I had a conversation with a layman about flying saucers ... I said "I don't think there are flying saucers." So my antagonist said, "Is it impossible that there are flying saucers? Can you prove it's impossible?" "No," I said, "I can't prove it's impossible. It's just very unlikely." At that he said, "You are very unscientific. If you can't prove it's impossible then how can you say it's unlikely?" ' (1965: 165)

25 Gish 1979: 40

26 Morris 1966: 114

27 They cite as their authority Karl Popper, but, like many others, they misinterpret him. Popper has characterized Darwinian evolutionary theory in three ways: as a historical theory, as a metaphysical doctrine, and as a research program. It is a historical theory because it provides an explanatory account of unique events – the descent of living forms on earth (see chapter 4, p. 86). It is not a universal law like Newton's law of gravitation. It is 'metaphysical' according to Popper's schema of demarcation between the empirical and the metaphysical *only* because it is not *as a whole* testable (though parts of it may be – for example, the thesis that modification proceeds by *small* steps), primarily because it cannot predict the next step in the evolution of a species (prediction is necessary for complete testability). And it is a research program, indeed the quintessential research program, since it has largely shaped biological research ever since it was first

promulgated. Even today, almost all biological research is done *within a Darwinian evolutionary framework.*

28 The arguments and counter-arguments about the fossil record, which are somewhat technical, are excellently summarized and discussed in Kitcher 1982: section 'Fear of Fossils,' pp. 106–17.

29 The 100th anniversary of Darwin's death

30 This is not, of course, to condone the use of illegal actions, such as breaking into, and defacing laboratories, and releasing experimental animals, as a means of protest, which has sometimes occurred; but that is a different issue.

31 Its full title is *Medical Nemesis: The Expropriation of Health* (1975).

32 Undertaken partly in the interest of improving technique and skills and partly for profit

33 That is, illness caused or exacerbated by doctors and hospitals

34 A group of fundamental particles including protons and neutrons

35 This was put more simply and without the hideous jargon of the Frankfurt School by the philosopher Alfred North Whitehead: 'Science ... has never cared to justify its truth or explain its meaning.' This is quite true, and it is one of the principal reasons for its success.

36 If this sounds far-fetched I refer the reader to a symposium on 'Witchcraft and the Epistemology of Science' held as part of the annual meeting of the British Association for the Advancement of Science in 1975, in the course of which a distinguished professor of sociology drew attention to 'the close similarities between the contemporary practice of science and the practice of witchcraft in tribal societies.' Also to the following passage from P. Wright, 'Astrology and Science in 17th Century England,' *Social Studies of Science* 5 (1975), 399: 'In recent years the task of distinguishing science from other cognitive systems has become much more complex. The traditional view that science is the polar opposite of such systems of thought as witchcraft, magic or astrology has been eroded almost to the point of extinction. On the one hand, historians and sociologists of science have been building up a view of scientific activity strikingly different from the hitherto accepted one; on the other, social anthropologists, in analysing so-called "alien belief-systems," have drawn attention to many features that appear to be far from alien to much scientific activity.'

37 See, for example, Grabner and Reiter's paper in Nowotny and Rose 1979, which is full of vulgar absurdities.

38 For an interesting treatment of *Naturphilosophie* in physics see Gower 1973: 301–56.

39 Mesmerism was the name given to the ideas of Franz Mesmer (1734–1815),

who developed a theory of 'animal magnetism'; phrenology, the theory that character traits and mental 'faculties' can be read from the topography of the skull, was the work of Francis Gall (1758–1828). An interesting book by Robert Darnton (*Mesmerism and the End of the Enlightenment in France*, Harvard University Press 1968) makes plain (whatever the author's intentions) that mesmerism, which spread its bounds in France far beyond Mesmer himself and became for a while a national hysteria, was the worst sort of quackery and contained many elements characteristic of anti-science.

40 Hegel also attacked Newton (in his *Philosophy of Nature*), yet he praised Kepler and believed that he had 'proved' Kepler's laws. The proof depends on absurd arguments, as the Reverend William Whewell made devastatingly clear in a paper read to the Cambridge Philosophical Society in 1849. It is easy to see why Hegel attacked Newton (for his materialism and mechanism), less easy to know why he praised Kepler; perhaps because Kepler 'returned to the Greeks' by uncovering the harmony of the spheres. Hegel betrayed a complete lack of understanding of Newton's reasoning, and his misguided criticisms undoubtedly stemmed from his attempt to reconstruct science a priori – for example, the criticism that Newton should have *proved* that a planet cannot move in a circle!

41 The quotations are taken from Gerlach 1973.

42 The reference is to Michael Oakeshott's 'The Voice of Poetry in the Conversation of Mankind' (1972).

43 The reference is to Richard Rorty, *Philosophy and the Mirror of Nature* (1980).

44 An example is Gunther Stent, a man who has made distinguished contributions to molecular biology, who – in startling repetition of Oswald Spengler – believes that 'the history of post-Renaissance science is now reaching its ironical denouement' (1978: 1).

References

Abrams, P. 1968. *The Origins of British Sociology: 1834–1914*. Chicago: University of Chicago Press

Asimov, I. 1977. 'Foreword: The Role of the Heretic,' in D. Goldsmith (ed.), *Scientists Confront Velikovsky*. Ithaca, NY: Cornell University Press

Barber, B. 1961. 'Resistance of Scientists to Scientific Discovery,' *Science* 134 (1 September): 596ff.

Barnes, B. 1974. *Scientific Knowledge and Sociological Theory*. London: Routledge and Kegan Paul

– 1982a. *T.S. Kuhn and Social Science*. London: Macmillan

– 1982b. 'The Science-Technology Relationship, a Model and a Query,' *Social Studies of Science* 12: 166ff.

Beck, W.S. 1961. *Modern Science and the Nature of Life*. Harmondsworth, England: Pelican Books

Ben-David, J. 1968. *Fundamental Research and the University*. Paris: Organisation for Economic Cooperation and Development

Blackburn, T.R. 1971. 'Sensuous-Intellectual Complementarity in Science,' *Science* 172 (4 June): 1003ff.

Block, N., and G. Dworkin (eds). 1977. *The I.Q. Controversy*. London: Quartet Books

Bok, S. 1984. *Secrets: On the Ethics of Concealment and Revelation*. New York: Vintage Books

Bondi, H. 1975. 'What Is Progress in Science?' in R. Harre (ed.), *Problems of Scientific Revolution*. Oxford: The Clarendon Press

Boslough, J. 1985. *Stephen Hawking's Universe*. New York: William Morrow

Brannigan, A. 1979. 'The Reification of Mendel,' *Social Studies of Science* 9: 423ff.

Broad, W. 1985. *Star Warriors*. New York: Simon and Schuster

Broad, W., and N. Wade. 1982. *Betrayers of the Truth: Fraud and Deceit in the Halls of Science*. New York: Simon and Schuster

Bronowski, J. 1973. *The Ascent of Man*. Boston: Little Brown

Brooks, H. 1967. 'Applied Science and Technological Progress,' *Science* 156: 1706ff.

– 1978. 'The Problems of Research Priorities,' *Daedalus* 107, no. 2: 171ff.

Bullock, A., and O. Stallybrass (eds). 1977. *The Harper Dictionary of Modern Thought*. Hagerstown, NY, and San Francisco: Harper and Row

Bunge, M. 1966. 'Technology as Applied Science,' *Technology and Culture* 8: 184ff.

Burt, C. 1975. *ESP and Psychology*. London: Weidenfeld and Nicolson

Burtt, E.A. 1932. *The Metaphysical Foundations of Modern Physical Science*. London: Routledge and Kegan Paul

Butterfield, H. 1965. *The Origins of Modern Science: 1300–1800*. New York: Macmillan and The Free Press

Cade, J. 1971. 'Aspects of Secrecy in Science,' *Impact of Science on Society* 21: 181ff.

Capra, F. 1975. *The Tao of Physics*. London: Wildwood House

Cardwell, D.S.L. 1973. 'Technology,' *Dictionary of the History of Ideas*, vol. 4. New York: Charles Scribner's Sons

Cassirer, E. 1950. *The Problem of Knowledge*. New Haven: Yale University Press

Cerullo, J.J. 1982. *The Secularisation of the Soul: Psychical Research in Modern Britain*. Philadelphia: Institute for the Study of Human Issues

Chedd, G. 1971. 'Romantic at Reason's Court' (an interview with Theodore Roszak), *New Scientist*, 4 March: 484ff.

Clarke, R. 1971. *The Science of War and Peace*. London: Jonathan Cape

Cloud, P. 1977. 'Scientific Creationism: A New Inquisition Brewing?' *The Humanist*, January–February issue

Cohen, I.B. 1976. 'Science and the Growth of the American Republic,' *Review of Politics* 38: 359ff.

Cohen, M.R. 1964. *Reason and Nature*. London: The Free Press, Collier-Macmillan

Cooter, R. 1980. 'Deploying Pseudo-science: Then and Now,' in M.P. Hanen, M.J. Osler, and R.G. Weyant (eds), *Science, Pseudo-science and Society*. Waterloo, Ontario: Wilfrid Laurier University Press

Cracraft, J. 1982. 'The Scientific Response to Creationism,' *Science Technology and Human Values* 7: 79ff.

Craig, G. 1983. 'Hitler without His Diaries,' *New York Review of Books*, 21 July: 4ff.

Crosland, M. (ed). 1975. *The Emergence of Science in Western Europe*. London: Macmillan
Daniels, G.H. 1971. *Science in American Society*. New York: Knopf
Dawkins, R. 1976. *The Selfish Gene*. Oxford: Oxford University Press
de Solla Price, D. 1962. *Science since Babylon*. New Haven: Yale University Press
– 1965. 'Is Technology Historically Independent of Science?' *Technology and Culture* 6: 553ff.
– 1969. 'The Structures of Publication in Science and Technology' in W. Gruber and G. Marquis (eds), *Factors in the Transfer of Technology*. Boston: MIT Press
d'Espagnat, B. 1979. 'The Quantum Theory and Reality,' *Scientific American* 241, no. 5 (November)
– 1980. Letter to the Editor, *Scientific American* 242, no. 5 (May)
Dewey, J. 1930. *The Quest for Certainty*. London: Allen and Unwin
– 1946. *The Public and Its Problems*. Chicago: University of Chicago Press
– 1948. *Reconstruction in Philosophy*. New York: Henry Holt
Dolby, R.G.A. 1975. 'What Can We Usefully Learn from the Velikovsky Affair?' *Social Studies of Science* 5: 165ff.
Edge, D.O., and J.N. Wolfe (eds). 1973. *Meaning and Control: Essays in Social Aspects of Science and Technology*. London: Tavistock Press
Eisenstein, M. 1977. 'Democratic Politics and Ideology,' *Canadian Journal of Political and Social Theory* 1: 98ff.
Farley, J. 1977. *The Spontaneous Generation Controversy from Descartes to Oparin*. Baltimore, Md: Johns Hopkins Press
Farrall, L.A. 1970. 'The Origins and Growth of the English Eugenics Movement.' PhD thesis, University of Indiana, Bloomington
Farrington, B. 1973. *Francis Bacon: Philosopher of Industrial Science*. London: Macmillan
Feyerabend, P. 1978a. *Against Method*. London: Verso Editions
– 1978b. *Science in a Free Society*. London: New Left Books
Feynman, R. 1965. *The Character of Physical Law*. London: The British Broadcasting Corporation
Forrest, D.W. 1974. *Francis Galton: The Life and Work of a Victorian Genius*. London: Elek Press
Franks, F. 1981. *Polywater*. Cambridge, Mass: MIT Press
Gale, G. 1981. 'The Anthropic Principle,' *Scientific American* 245, no. 6 (December): 154ff.
Gardner, M. 1957. *Fads and Fallacies: In the Name of Science*. New York: Dover Publications
– 1981. *Science: Good, Bad and Bogus*. Buffalo: Prometheus Books

– 1985. 'Physics: The End of the Road?' *New York Review of Books* 32, no. 10 (13 June): 31–4
Gasman, D. 1971. *The Scientific Origins of National Socialism: Social Darwinism in Ernst Haeckel and the German Monist League*. New York: Neale Watson Academic Publications
Gellner, E. 1974. *The Devil in Modern Philosophy*. London: Routledge and Kegan Paul
George, W. 1982. *Darwin*. Modern Masters Series; Glasgow: William Collins
Gerlach, W. 1973. 'Goethe as a Scientist,' *Times Literary Supplement*, 3 August: 907ff.
Ghiselin, M.T. 1969. *The Triumph of the Darwinian Method*. Berkeley, Calif.: University of California Press
Gibbons, M., and C. Johnson. 1970. 'Relationship between Science and Technology,' *Nature* 227, 11 July: 125ff.
Gillispie, C.C. 1960. *The Edge of Objectivity*. Princeton: Princeton University Press
Gish, D. 1979. *Evolution: The Fossils Say No!* San Diego: Creation-Life Publishers
Goldsmith, D. 1977. *Scientists Confront Velikovsky*. Ithaca, NY: Cornell University Press
Goodell, R., and J. Goodfield. 1978. 'Rorvik's Baby,' *The Sciences*, September
Goodfield, J. 1981. *An Imagined World: A Story of Scientific Discovery*. New York: Harper and Row
Gordon, H. 1987. *Extrasensory Deception*. Buffalo, NY: Prometheus Books
Gould, S.J. 1977. *Ever since Darwin*. New York: W.W. Norton Company
– 1978. 'Morton's Ranking of Races by Cranial Capacity,' *Science* 200: 503ff.
– 1980a. *The Panda's Thumb*. New York: W.W. Norton Company
– 1980b. 'Jensen's Last Stand,' *New York Review of Books*, 1 May
– 1981a. *The Mismeasure of Man*. New York: W.W. Norton Company
– 1981b. 'Ice-Nine, Russian Style,' *New York Times Sunday Book Review*, 30 August
– 1983. *Hens' Teeth and Horses' Toes*. New York: W.W. Norton Company
Gower, B. 1973. 'Speculation in Physics: The History and Practice of *Naturphilosophie*,' *Studies in the History and Philosophy of Science* 3, no. 4: 301–56
Haberer, J. 1972. 'Politicisation of Science,' *Science* 178 (17 November): 713–23
Habermas, J. 1970. *Toward a Rational Society*. Boston: Beacon Press
– 1976. 'The Analytical Theory of Science and Dialectics,' in T.W. Adorno, et al., *The Positivist Dispute in German Sociology*. London: Heinemann
Haller, J.S. 1971. *Outcasts from Evolution: Scientific Attitudes to Racial Inferiority 1859–1900*. Urbana: University of Illinois Press

Hammond, M. 1982. 'The Expulsion of the Neanderthals from Human Ancestry,' *Social Studies of Science* 12: 1ff.

Hannay, N.B., and R.E. McGinn. 1980. 'The Anatomy of Modern Technology,' *Daedalus* 109, no. 1, Winter: 25–53

Hansel, C.E.M. 1980. *ESP and Parapsychology*. Buffalo, NY: Prometheus Books

Hardin, C.L. 1981. 'Table-Turning, Parapsychology, and Fraud,' *Social Studies of Science* 11: 249ff.

Hearnshaw, L.S. 1979. *Cyril Burt: Psychologist*. London: Hutchinson

Heims, S.J. 1980. *John von Neumann and Norbert Wiener: From Mathematics to the Technologies of Life and Death*. Cambridge, Mass.: MIT Press

Heisenberg, W. 1958. 'The Representation of Nature in Contemporary Physics,' *Daedalus* 87, Summer: 95ff.

Hines, T. 1987. *Pseudo-science and the Paranormal*. Buffalo, NY: Prometheus Books

Hixon, J. 1975. *The Patchwork Mouse*. New York: Anchor Press

Hoch, P.K. 1987. 'Migration and the Generation of New Scientific Ideas,' *Minerva* 25, no. 3: 209–37

Hoddeson, L.H. 1980. 'The Entry of the Quantum Theory of Solids into the Bell Telephone Laboratories 1925–1940,' *Minerva* 18, no. 3: 422–47

Hofstadter, R. 1955. *Social Darwinism in American Thought*. Boston: Beacon Press

Holton, G. 1973. 'The Roots of Complementarity,' in his *Thematic Origins of Scientific Thought*. Cambridge, Mass.: Harvard University Press

Hull, D. 1973. *Darwin and His Critics*. Chicago: University of Chicago Press

Husserl, E. 1970. *The Crisis of the European Sciences and Transcendental Phenomenology*. Evanston, Ill.: Northwestern University Press

Jacob, F. 1982. *The Possible and the Actual*. New York: Pantheon Books

Jammer, M. 1975. *The Philosophy of Quantum Mechanics*. Cambridge, Mass.: Harvard University Press

Jensen, A. 1969. 'How Much Can We Boost I.Q. and Scholastic Achievement?' *Harvard Educational Review* 39: 1ff.

– 1972. *Genetics and Education*. New York: Harper and Row

– 1979. *Bias in Mental Testing*. New York: Free Press

Joravsky, D. 1970. *The Lysenko Affair*. Cambridge, Mass.: Harvard University Press

Judson, H.F. 1980. *The Search for Solutions*. New York: Holt Rinehart

Juergens, R. 1966. 'Minds in Chaos,' in A. de Grazia et al. (eds), *The Velikovsky Affair: Scientism versus Science*. New Hyde Park, NY: University Books

Kamin, L.J. 1974. *The Science and Politics of I.Q.* Potomac, Md: Lawrence Erlbaum

Keller, A. 1984. 'Has Science Created Technology?' *Minerva* 22, no. 2: 160–82

Keller, E. 1983. *A Feeling for the Organism: The Life and Work of Barbara McClintock*. San Francisco: W.H. Freeman and Company
Kevles, D.J. 1984. *In the Name of Genetics*. New York: Alfred Knopf
Keynes, G. 1966. *The Life of William Harvey*. Oxford: The Clarendon Press
Keynes, J.M. 1972. *Essays in Biography*, vol. 9, *Collected Writings*. London: Macmillan/St Martins Press, for the Royal Economic Society
Kimmelman, B.A. 1983. 'The American Breeders' Association: Genetics and Eugenics in an Agricultural Context 1903–1913,' *Social Studies of Science* 13: 163ff.
Kitcher, P. 1982. *Abusing Science: The Case against Creationism*. Cambridge, Mass.: MIT Press
Klaw, S. 1968. *The New Brahmins*. New York: William Morrow
Klotz, I.M. 1980. 'The N-Ray Affair,' *Scientific American* 242, no. 5 (May): 168ff.
Kneen, P. 1984. *Soviet Scientists and the State*. London: Macmillan
Knight, D.M. 1975. 'German Science in the Romantic Period,' in Crosland (1975)
Koestler, A. 1964. *The Sleepwalkers*. Harmondsworth, England: Penguin Books
– 1971. *The Case of the Midwife Toad*. London: Hutchinson
Kuhn, T.S. 1962. *The Structure of Scientific Revolutions*. Chicago: University of Chicago Press
– 1970. 'Logic of Discovery or Psychology of Research?' in I. Lakatos and A. Musgrave (eds). *Criticism and the Growth of Knowledge*. Cambridge: Cambridge University Press
– 1977. *The Essential Tension*. Chicago: University of Chicago Press
Kuhner, H. 1984. 'Between Autonomy and Planning,' *Minerva* 22: 13ff.
Langrish, J., et al. 1972. *Wealth from Knowledge*. London: Macmillan
Layzer, D. 1977. 'Science or Superstition? A Physical Scientist Looks at the I.Q. Controversy,' in Block and Dworkin (1977)
Lewontin, R. 1977. 'Further Remarks on Race and the Genetics of Intelligence,' in Block and Dworkin (1977)
– 1981. 'The Inferiority Complex,' *New York Review of Books*, 22 October
Lubrano, L. 1981. 'National and International Politics in US-USSR Scientific Cooperation,' *Social Studies of Science* 11: 451ff.
Lumsden, C., and E.O. Wilson. 1981. *Genes, Mind and Culture*. Cambridge, Mass.: Harvard University Press
– 1983. *Promethean Fire: Reflections on the Origin of Mind*. Cambridge, Mass.: Harvard University Press
Luria, S. 1984. *A Slot Machine, a Broken Test Tube: An Autobiography*. New York: Harper and Row
Macdonald, S. 1986. 'Controlling the Flow of High Technology Information from the United States to the Soviet Union,' *Minerva* 24, no. 1: 39–73

Mackenzie, D. 1976. 'Eugenics in Britain,' *Social Studies of Science* 6: 499ff.
– 1981. 'Eugenics and the Rise of Mathematical Statistics in Britain,' in J. Irvine et al., *Demystifying Social Statistics*. London: Pluto Press
Magyar, G. 1977. 'Pseudo-effects in Experimental Physics: Some Notes for Case Studies,' *Social Studies of Science* 7: 242ff.
Malitza, M. 1979. 'Transdisciplinary Communication,' in T. Segestedt (ed), *Ethics for Science Policy*. Oxford: Pergamon Press
Marcuse, H. 1966. *One-Dimensional Man*. Boston: Beacon Press
Marks, D., and R. Kammann. 1980. *The Psychology of the Psychic*. Buffalo: Prometheus Books
Marwick, B. 1978. 'The Soal-Goldney Experiments with Basil Shackleton: New Evidence of Data Manipulation,' *Proceedings of the Society for Psychical Research* 56: 250ff.
Mauskopf, S.H., and M.R. McVaugh. 1980. *The Elusive Science: Origins of Experimental Psychical Research*. Baltimore, Md: Johns Hopkins University Press
Maynard-Smith, J. 1982. 'Genes and Memes,' *London Review of Books*, 4–18 February: 3ff.
Mayr, E. 1982. *The Growth of Biological Thought*. Cambridge, Mass.: Harvard University Press
Mayr, O. 1976. 'The Science-Technology Relationship as a Historiographic Problem,' *Technology and Culture* 17: 663ff.
McCloskey, M. 1983. 'Intuitive Physics,' *Scientific American* 248, no. 4 (April)
McGowan, C. 1984. *In the Beginning: A Scientist Shows Why the Creationists Are Wrong*. Buffalo: Prometheus Books
Medawar, P. 1967. *The Art of the Soluble*. London: Methuen
– 1969. *Induction and Intuition in Scientific Thought*. London: Methuen
– 1972. *The Hope of Progress*. London: Methuen
– 1977a. 'Unnatural Science,' *New York Review of Books*, 23 February
– 1977b. *The Life Sciences*. London: Wildwood House
Medvedev, Z. 1969. *The Rise and Fall of T.D. Lysenko*. New York: Columbia University Press
– 1971. *The Medvedev Papers*. London: Macmillan
– 1978. *Soviet Science*. New York: W.W. Norton Company
Merton, R. 1973. 'The Normative Structure of Science,' in his *The Sociology of Science*, 2nd ed. Chicago: University of Chicago Press
Moore, J.A. 1976. 'Creationism in California,' in G. Holton and W. Blanpied (eds), *Science and Its Public: The Changing Relationship*. Dordrecht, Holland: D. Reidel Publishing Company
Morris, H. 1966. *Studies in the Bible and Science*. Philadelphia: Presbyterian and Reformed Publishing Company

Mulkay, M. 1970. 'Knowledge and Utility,' *Social Studies of Science* 9: 69ff.
– 1979. *Science and the Sociology of Knowledge*. London: Allen and Unwin
Nelkin, D. 1977a. 'Creationism versus Evolution,' in E. Mendelsohn et al. (eds), *The Social Production of Scientific Knowledge*. Dordrecht, Holland: Reidel Company
– 1977b. *Science Textbook Controversies and the Politics of Equal Time*. Cambridge, Mass.: MIT Press
– 1982. *The Creation Controversy: Science or Scripture in the Schools*. New York: W.W. Norton Company
Newell, N.D. 1982. *Creation and Evolution: Myth or Reality*. New York: Columbia University Press
Norton, B. 1978a. 'Karl Pearson and Statistics,' *Social Studies of Science* 8: 3ff.
– 1978b. Article in *New Scientist*, 24 April: 223
Nowotny, H., and H. Rose (eds). 1979. *Counter-Movements in the Sciences*. Dordrecht, Holland: Reidel Company
Oakeshott, M. 1972. 'The Voice of Poetry in the Conversation of Mankind,' in his *Rationalism and Politics*. London: Methuen
Oppenheim, J. 1985. *Spiritualism and Psychical Research in England 1850–1914*. Cambridge: Cambridge University Press
Orel, V.O. 1984. *Mendel*. Oxford and New York: Oxford University Press
Pagels, H.R. 1985. *Perfect Symmetry: The Search for the Beginning of Time*. New York: Simon and Schuster
Pais, A. 1982. *Subtle Is the Lord: The Science and Life of Albert Einstein*. Oxford: Oxford University Press
Paul, H.W. 1979. *The Edge of Contingency: French Catholic Reaction to Scientific Change from Darwin to Duhem*. Gainesville, Fla.: University of Florida Press
Pinch, T. 1979. 'Normal Explanations of the Paranormal,' *Social Studies of Science*, 9: 329ff.
Pinch, T., and H.M. Collins. 1981. 'The Construction of the Paranormal,' in Collins, H.M. (ed.). *Sociology of Scientific Knowledge: A Source Book*. Bath: Bath University Press
Polanyi, M. 1946. *Science, Faith and Society*. London: Oxford University Press
– 1951. *The Logic of Liberty*. London: Routledge and Kegan Paul
– 1958. *Personal Knowledge*. New York: Harper and Row
– 1962. 'The Republic of Science,' *Minerva* 1: 54ff.
– 1969. 'The Growth of Science in Society,' in his *Knowing and Being*. London: Routledge and Kegan Paul
Popper, K. 1962. *Conjectures and Refutations*. New York: Basic Books
– 1966. 'Quantum Mechanics without the Observer,' in M. Bunge (ed), *Quantum Theory and Reality*. Berlin: Springer-Verlag

– 1969. *The Open Society and Its Enemies*. London: Routledge and Kegan Paul (reprint of 1945)
– 1972. *Objective Knowledge*. Oxford: Clarendon Press
– 1982. *Quantum Theory and the Schism in Physics*. London: Hutchinson
– 1983. *Realism and the Aim of Science*. London: Hutchinson
Rasmussen, D. 1985. 'The Revolution in Not-So-Conventional Weapons,' *Science for the People* 17, no. 1–2: 12–15
Redondi, P. 1987. *Galileo Heretic*. Princeton: Princeton University Press
Rhinelander, P. 1973. *Is Man Incomprehensible to Man?* San Francisco: W.H. Freeman Company
Ronan, C., and J. Needham. 1978. *The Shorter Science and Civilisation in China*. Cambridge: Cambridge University Press
Rorty, R. 1980. *Philosophy and the Mirror of Nature*. Princeton: Princeton University Press; Oxford: Basil Blackwell
Rose, H. 1979. 'Hyper-Reflexivity: A New Danger for the Counter-Movements,' in Nowotny and Rose (1979)
Rose, S. 1976. 'Scientific Racism and Ideology: The I.Q. Racket from Galton to Jensen,' in S. Rose and H. Rose (eds), *The Political Economy of Science*. London: Macmillan
Ross, S. 1962. 'Scientist: The Story of a Word,' *Annals of Science* 18, no. 2: 65–85
Rossi, P. 1973. 'Baconianism,' *Dictionary of the History of Ideas*, vol. 4, New York: Charles Scribner's Sons
Rostand, J. 1960. *Error and Deception in Science*. London: Hutchinson
Roszak, T. 1968. *The Making of a Counter-Culture*. Garden City, NY: Doubleday
– 1972. *Where the Wasteland Ends*. Garden City, NY: Doubleday
Roy, R. 1984. 'Alternatives to Review by Peers: A Contribution to the Theory of Scientific Choice,' *Minerva* 22: 316ff.
Ruse, M. 1982. *Darwin Defended: A Guide to the Evolution Controversies*. Reading, Mass.: Addison-Wesley
Russell, B. 1950. *Unpopular Essays*. London: Allen and Unwin
Sagan, C. 1979. 'Venus and Dr. Velikovsky,' in his *Broca's Brain*. New York: Random House
Sapolsky, H. 1977. 'Science, Technology and Military Policy,' in I. Spiegel-Rosing and D. de Solla Price (eds), *Science, Technology and Society*. London and Beverly Hills, Calif.: Sage Publications
Sayre, A. 1975. *Rosalind Franklin and D.N.A.* New York: Norton
Schafer, W. (ed). 1983. *Finalization in Science: The Social Orientation of Scientific Progress*. Dordrecht, Holland: Reidel Company

Schilpp, P.A. (ed). 1974. *The Philosophy of Karl Popper.* The Library of Living Philosophers, vol. 14; La Salle, Ill.: The Open Court Press

Schmookler, J. 1966. *Invention and Economic Growth.* Cambridge, Mass.: Harvard University Press

Schroeder-Gudehus, B. 1973. 'Challenge to Trans-National Loyalties,' *Science Studies* 3: 93ff.

Schultz, T. 1980. 'The Productivity of Research – The Politics and Economics of Research,' *Minerva* 18: 644–51.

Shapiro, B. 1969. 'Law and Science in Seventeenth Century England,' *Stanford Law Review* 21: 727ff.

Shattock, M. 1987. 'The Last Days of the University Grants Committee,' *Minerva* 25, no. 4: 471–85

Shattuck, J. 1984. 'Federal Restrictions on the Free Flow of Academic Information and Ideas,' *Minerva* 22: 424ff.

Shils, E. 1984. 'Secrecy and Freedom of Communication in American Science,' *Minerva* 22: 421ff.

Simon, H. 1968. *The Sciences of the Artificial.* Cambridge, Mass.: MIT Press

Skolimowski, H. 1966. 'The Structure of Thinking in Technology,' *Technology and Culture* 7: 371ff.

Snow, C.P. 1959. *The Two Cultures and the Scientific Revolution.* Cambridge: Cambridge University Press

Stecchini, L. 1966. 'Astronomical Theory and Historical Data,' in A. de Grazia, et al. (eds). *The Velikovsky Affair.* New York: University Books

Stent, G. 1978. *Paradoxes of Progress.* San Francisco: W.H. Freeman and Company

Stepan, N. 1982. *The Idea of Race in Science: Great Britain 1800–1960.* London: Macmillan

Sullivan, J.W.N. 1933, 1938. *The Limitations of Science.* Harmondsworth, England: Penguin Books

Sutton, J.R. 1984. 'Organizational Autonomy and Professional Norms in Science: A Case Study of Lawrence Livermore Laboratory,' *Social Studies of Science* 14, no. 1: 197–224

Thomas, L. 1979. *The Medusa and the Snail.* New York: The Viking Press

Trigg, R. 1980. *Reality at Risk: A Defence of Realism in Philosophy and the Sciences.* Sussex: The Harvester Press: Totowa, NJ: Barnes and Noble

Van Helden, A. 1977. 'The Invention of the Telescope,' *Transactions of the American Philosophical Society* 67, pt. 4

Velikovsky, I. 1967. *Worlds in Collision.* New York: Dell Paperbacks

– 1976. 'Are the Moon's Scars Only Three Thousand Years Old?' in Editors of Pensée, *Velikovsky Reconsidered.* New York: Doubleday

– 1983. *Stargazers and Gravediggers: Memoirs to Worlds in Collision*. New York: William Morrow

Weisskopf, V. 1980. Letter to the Editor, *Scientific American* 242, no. 5 (May)

Wheeler, J.A. 1973. *Gravitation*. San Francisco: W.H. Freeman Company

– 1975. 'The Universe as Home for Man,' in O. Gingerich (ed), *The Nature of Scientific Discovery*. Washington, DC: The Smithsonian Institution Press

Wigner, E. 1967. *Symmetries and Reflections*. Cambridge, Mass.: MIT Press

Will, C.M. 1986. *Was Einstein Right?: Putting General Relativity to the Test*. New York: Basic Books

Wilson, E.O. 1975. *Sociobiology: The New Synthesis*. Cambridge, Mass.: Harvard University Press

– 1977. 'Biology and the Social Sciences,' *Daedalus* 106: 127–40

– 1978. *On Human Nature*. Cambridge, Mass.: Harvard University Press

Wysong, R.L. 1976. *The Creation-Evolution Controversy*. Midland, Me: Inquiry Press

York, D. 1981. Article in the *Globe and Mail* (Toronto), 16 February

Young, R.M. 1985. 'The Impact of Darwin on Conventional Thought,' in his *Darwin's Metaphor: Nature's Place in Victorian Culture*. Cambridge: Cambridge University Press

Ziman, J. 1968. *Public Knowledge: The Social Dimension of Science*. Cambridge: Cambridge University Press

– 1976. *The Force of Knowledge: The Scientific Dimension of Science*. Cambridge: Cambridge University Press

– 1978a. *Reliable Knowledge: An Exploration of the Grounds for Belief in Science*. Cambridge: Cambridge University Press

– 1978b. 'Solidarity within the Republic of Science,' *Minerva* 16: 4ff.

– 1981. 'What Are the Options?: Social Determinants of Personal Research Plans,' *Minerva* 19: 1ff.

– 1987. 'The Problem of Problem Choice,' *Minerva* 25, no. 1–2: 92–106

Index